Springer Series on

ATOMIC, OPTICAL, AND PLASMA PHYSICS 53

Springer Series on

ATOMIC, OPTICAL, AND PLASMA PHYSICS

The Springer Series on Atomic, Optical, and Plasma Physics covers in a comprehensive manner theory and experiment in the entire field of atoms and molecules and their interaction with electromagnetic radiation. Books in the series provide a rich source of new ideas and techniques with wide applications in fields such as chemistry, materials science, astrophysics, surface science, plasma technology, advanced optics, aeronomy, and engineering. Laser physics is a particular connecting theme that has provided much of the continuing impetus for new developments in the field. The purpose of the series is to cover the gap between standard undergraduate textbooks and the research literature with emphasis on the fundamental ideas, methods, techniques, and results in the field.

47 **Semiclassical Dynamics and Relaxation**
By D.S.F. Crothers

48 **Theoretical Femtosecond Physics**
Atoms and Molecules in Strong Laser Fields
By F. Großmann

49 **Relativistic Collisions of Structured Atomic Particles**
By A. Voitkiv and J. Ullrich

50 **Cathodic Arcs**
From Fractal Spots to Energetic Condensation
By A. Anders

51 **Reference Data on Atomic Physics and Atomic Processes**
By B.M. Smirnov

52 **Relativistic Transitions in the Hydrogenic Atoms**
Elementary Theory
By R. Boudet

53 **The Kinetic Theory of a Dilute Ionized Plasma**
By L.S. García-Colín and L. Dagdug

For other titles published in this series, go to
www.springer.com/series/411

Leopoldo S. García-Colín
Leonardo Dagdug

The Kinetic Theory of a Dilute Ionized Plasma

Springer

Leopoldo S. García-Colín
Universidad Autónoma
Metropolitana-Iztapalapa
Depto. Física
Av. San Rafael Atlixco 186
09340 México, D.F.
México

Leonardo Dagdug
Universidad Autónoma
Metropolitana-Iztapalapa
Depto. Física
Av. San Rafael Atlixco 186
09340 México, D.F.
México

Springer Series on Atomic, Optical, and Plasma Physics ISSN 1615-5653

ISBN 978-90-481-8109-4 e-ISBN 978-1-4020-9330-2

Printed on acid-free paper.

9 8 7 6 5 4 3 2 1

springer.com

Acknowledgement

The authors are indebted to Alfredo Sandoval, Ana Laura García Perciante, and Valdemar Moratto for a careful reading of the manuscript, and for a number of useful suggestions which have been incorporated into it.

Contents

Introduction 1

Part I Vector Transport Processes

1 Non-equilibrium Thermodynamics 5

2 The Problem 13
2.1 Conservation Equations 14
2.2 The H Theorem and Local Equilibrium 18

3 Solution of the Boltzmann Equation 25

4 Calculation of the Currents 41
4.1 Diffusion Effects 41
4.2 Flow of Heat 45

5 Solution of the Integral Equations 51

6 The Transport Coefficients 61

7 Discussion of the Results 73

Part II Tensorial Transport Processes

8 Viscomagnetism 83
8.1 The Integral Equation 83
8.2 The Stress Tensor 93
8.3 The Integral Equation 99
8.4 Comparison with Thermodynamics 102

9 Magnetohydrodynamics 107

Appendix A Calculation of M 125

Appendix B Linearized Boltzmann Collision Kernels 129

Appendix C The Case when $\vec{B} = \vec{0}$ 133

Appendix D The Collision Integrals 145

Appendix E Calculation of the Coefficients $a_i^{(0)}$, $a_i^{(1)}$, $d_i^{(0)}$ and $d_i^{(1)}$ 153

Appendix F 155

Appendix G 157

Appendix H 159

Appendix I List of Marshall's Equations and Notation 161
I.1 Equations 161
I.2 Notation 162

Index 165

Introduction

The contents of this book are the result of work performed in the past three years to provide some answers to questions raised by several colleagues working in astrophysics. Examining several transport processes in plasmas related to dissipative effects in phenomena such as cooling flows, propagation of sound waves, thermal conduction in the presence of magnetic fields, angular momentum transfer in accretion disks, among many, one finds a rather common pattern. Indeed when values for transport coefficients are required the overwhelming majority of authors refer to the classical results obtained by L. Spitzer and S. Braginski over forty years ago. Further, it is also often mentioned that under the prescribed working conditions the values of such coefficients are usually insufficient to provide agreement with observations.

The methodology followed by these authors is based upon Landau's pioneering idea that collisions in plasmas may be substantially accounted for when viewed as a diffusive process. Consequently the ensuing basic kinetic equation is the Fokker-Planck version of Boltzmann's equation as essentially proposed by Landau himself nearly 70 years ago. Curiously enough the magnificent work of the late R. Balescu in both Classical and Non-Classical transport in plasmas published in 1988 and also based on the Fokker-Planck equation is hardly known in the astrophysical audience. The previous work of Spitzer and Braginski is analyzed with much more rigorous vision in his two books on the subject.

With this background in hand the question that came to our minds is why, if true, the full Boltzmann's equation had never been used in dealing at least with the kinetic theory of dilute plasmas. In their well known and comprehensive treatment on the kinetic theory of non-uniform gases, Chapman and Cowling never developed the theory as they did with ordinary gases. A further attempt was made in 1960 by W. Marshall in three unpublished reports issued by the Harwell Atomic Energy Establishment in

L.S. García-Colín, L. Dagdug, *The Kinetic Theory of a Dilute Ionized Plasma*, Springer Series on Atomic, Optical, and Plasma Physics 53

Harwell, England. And also, none of all the authors in this field with the sole exception of Balescu who did it partially, took the kinetic equation of their choice to provide the microscopic basis of linear irreversible thermodynamics therefore, providing, among many other results, a microscopic basis of magnetohydrodynamics.

This is the main objective of this book. Starting from the full Boltzmann equation for an inert dilute plasma and using the Hilbert-Chapman-Enskog method to solve the first two approximations in Knudsen's parameter we construct all the transport properties of the system within the framework of linear irreversible thermodynamics. This includes a systematic study of all possible cross effects which except for a few cases dealt with by Balescu, today to our knowledge, have never been mentioned in the literature. The equations of magnetohydrodynamics, including the rather surprising results here obtained for the viscomagnetic effects, for dilute plasmas may be then fully assessed. We expect that this material will thus be useful to graduate students and researchers involved in work with non-confined plasmas specially in astrophysical problems.

July 2008

L.S. García-Colín
L. Dagdug

Part I

Vector Transport Processes

Chapter 1

Non-equilibrium Thermodynamics

The main objective of this book is to place the kinetic theory of a dilute plasma within the tenets of what is known as Classical (Linear) Irreversible Thermodynamics (CIT). Since this subject is quite often beyond the average knowledge of the younger generation of physicists and physical chemists we feel that it is useful to give a brief review of its basic concepts so that the reader appreciates better how and why we are seeking the results to be presented in the main text.

CIT, being a phenomenological theory is based essentially on four basic assumptions, namely,

1. The local equilibrium assumption (LEA)

2. The validity of the conservation equations

3. The linear constitutive equations and positive definiteness of the uncompensated heat (entropy production)

4. Onsagers' reciprocity theorem

In what follows we shall discuss as thoroughly as possible the basic ideas behind each assumption, leaving the reader to pursue more details in the standard texts on the subject [1]-[7]. Let us start with the LEA. Consider any arbitrary system which is not in thermodynamic equilibrium. For purely didactical reasons the reader may think of a fluid enclosed in a volume V.

L.S. García-Colín, L. Dagdug, *The Kinetic Theory of a Dilute Ionized Plasma*,
Springer Series on Atomic, Optical, and Plasma Physics 53

Let us now partition this volume in small cells such that the number of particles in each cell with coordinates $\vec{r}$, $\vec{r} + d\vec{r}$ at time t contains enough particles to be considered as a continuum but small compared with the total number of particles in the system, say N. The LEA asserts that within each cell a thermodynamic equilibrium state prevails. For instance, if $n(\vec{r}, t)$ is the particle density in the cell characterized by its position $\vec{r}$ at time t and $T(\vec{r}, t)$ the temperature inside the cell, any other thermodynamic quantity, for instance the entropy $s(\vec{r}, t)$ will be related to $n(\vec{r}, t)$ and $T(\vec{r}, t)$ as

$$s(\vec{r}, t) = s[n(\vec{r}, t), T(\vec{r}, t)] \tag{1.1}$$

precisely by the same relationship that holds for these variables in the equilibrium state. The local equation of state for an ideal gas would read

$$p(\vec{r}, t) = n(\vec{r}, t) k_B T(\vec{r}, t) \tag{1.2}$$

k_B being Boltzmann's constant. And so on.

These equations bring us in a natural way to the second assumption. Think of a monatomic fluid for the moment in the absence of sources and sinks. If we chose to describe the states of this fluid by the "natural" variables, the local particles density $n(\vec{r}, t)$, the local hydrodynamic velocity $\vec{u}(\vec{r}, t)$ (or $m\vec{u}(\vec{r}, t)$ its momentum) and the local energy density $e(\vec{r}, t)$ these variables will satisfy clearly, conservation equations. Use of this fact and Eq. (1.1) with $e(\vec{r}, t)$ instead of $T(\vec{r}, t)$ plus the standard techniques of ordinary calculus lead us in a straightforward fashion to an equation describing the evolution of the local entropy $s(\vec{r}, t)$. In fact if $\rho(\vec{r}, t) = mn(\vec{r}, t)$, m being the mass of the particles, one finds that,

$$\frac{\partial(\rho s)}{\partial t} + \text{div } \vec{J}_s = \sigma \tag{1.3}$$

which is a balance type equation for ρs. $\vec{J}_s$, the entropy flux, gives the amount of entropy flowing through the boundaries to the system and σ, the uncompensated heat or entropy production, measures the entropy generated inside the system due to the dissipative processes. Its existence goes back to Clausius who indeed identified it with the uncompensated heat which should arise from dissipation. Its analytical expression was first identified by T. de Donder in chemical reactions and later brought into its present form by Meixner. Indeed, in the derivation of Eq. (1.3) one finds that

$$\sigma = \sum \overleftrightarrow{J}_i \odot \overleftrightarrow{X}_i = -\frac{\vec{J}_q}{T} \cdot \text{grad } T - \frac{1}{T} \overset{\circ}{\overleftrightarrow{\tau}} : (\overset{\circ}{\text{grad}}\ \vec{u})^s - \frac{1}{T} \tau \text{div } \vec{u} \tag{1.4}$$

where $\overleftrightarrow{J}_i$ and $\overleftrightarrow{X}_i$ denote the fluxes and their corresponding forces respectively, and $\odot$ the contraction of tensors of equal rank. The second equality illustrates its nature for an ordinary monatomic fluid. $\vec{J}_q$ is the heat flow vector and the momentum flow $\overleftrightarrow{\tau}$ is split into its symmetric traceless part $\overset{\circ}{\overleftrightarrow{\tau}}$ and its trace τ. Eq. (1.4) clearly fulfills Clausius' predictions.

These results bring us to the third assumption. The conservation equations for a monatomic fluid are the set of five differential equations for the state variables ρ, $\vec{u}$ and e but contain fourteen unknowns, these variables plus the three components of $\vec{J}_q$ plus the six independent components of the stress tensor $\overleftrightarrow{\tau}$ assumed to be symmetric. We thus need nine additional equations to express $\vec{J}_q$ and $\overleftrightarrow{\tau}$ in terms of the independent variables. Notice that $T(\vec{r},t)$ may be introduced through the LEA since $e(\vec{r},t) = e(n(\vec{r},t), T(\vec{r},t))$. These additional equations known in the literature as the "constitutive equations" are completely foreign to thermodynamics. They may be extracted from experiment or from a microscopic theory. If we now assume (assumption 3) that the relationship between fluxes and forces is linear so that in general,

$$\overleftrightarrow{J}_i = \sum L_{ik} \overleftrightarrow{X}_k, \tag{1.5}$$

we may obtain a complete set for the time evolution equations of the local state variables. For a monatomic fluid, Eq. (1.5) reduces to

$$\vec{J}_q = -\kappa \text{grad } T \quad \text{Fourier} \tag{1.6}$$

$$\left.\begin{aligned} \overset{\circ}{\overleftrightarrow{\tau}}{}^{s} &= -\eta (\overset{\circ}{\text{grad}}\ \vec{u})^s \\ \tau &= -\zeta \text{div } \vec{u} \end{aligned}\right\} \quad \text{Naviere-Newton}$$

As it is shown in any standard text on the subject, when Eqs. (1.6) are substituted into the conservation equations one gets a set of non-linear, second order in space, first order in time differential equations for n, $\vec{u}$ and T known as the Navier-Stokes-Fourier equations of hydrodynamics. These equations require the knowledge of a local equation of state (c.f. Eq. (1.1)), of the transport coefficients κ, the thermal conductivity, η, the shear viscosity, and ζ, the bulk viscosity in addition to well defined boundary and initial conditions to seek for a solution.

In spite of its centennial age these equations still pose immense problems to mathematical physicists and hydrodynamicists in finding stable solutions [8].

There is however an additional feature brought by the assumption written in Eq. (1.6). For the case of a fluid when substituted into Eq. (1.4) yields

$$\sigma = \frac{\kappa}{T^2}(\text{grad } T)^2 + \frac{\eta}{T}(\overset{\circ}{\text{grad }} \vec{u})^s : (\overset{\circ}{\text{grad }} \vec{u})^s + \frac{\zeta}{T}(\text{div } T)^2, \tag{1.7}$$

a quadratic form for σ which is positive definite if κ, η and ζ are positive, a fact drawn from experiment. This means that

$$\sigma > 0 \tag{1.8}$$

a quite strong statement implying complete consistency with the second law of thermodynamics. Its extension to open systems is not trivial but we shall not discuss this here, it is treated in references [3] and [6].

There is an additional feature about Eq. (1.5) which deserves mention before going into the last assumption behind CIT. Why a summation in this equation is necessary. This occurs in systems where two or more thermodynamics forces are present and are of the same tensorial rank. There are many examples of this nature in many physical systems, let us mention here two of the most frequent ones. Suppose we have a mixture of two monatomic fluids subject to the same temperature gradient and to a concentration gradient of one of the species ($c_1 + c_2 = 1$) so only one concentration is independent. This implies that the total heat flow $\vec{J}_q$ and the mass flux $\vec{J}_m$ are

$$\begin{aligned} \vec{J}_q &= -L_{qq}\text{grad } T - L_{qm}\text{grad } c_1 \\ \vec{J}_m &= -L_{mq}\text{grad } T - L_{mm}\text{grad } c_1 \end{aligned} \tag{1.9}$$

A concentration gradient may give rise to a heat flow an effect called thermal diffusion or Dufour effect, whereas a thermal gradient gives rise to a mass flux or diffusion termo effect (Soret effect). These "cross effects" are very important in multicomponent systems. In the case of a conducting solid subject to both thermal gradients and an electric field one has that ($\vec{E} = -\text{grad } \phi$)

$$\begin{aligned} \vec{J}_q &= -L_{qq}\text{grad } T - L_{qe}\text{grad } \phi \\ \vec{J}_e &= -L_{eq}\text{grad } T - L_{ee}\text{grad } \phi \end{aligned} \tag{1.10}$$

where $\vec{J}_e$ is the charge flux or electrical current. L_{eq} is known as the electrophoresis or Benedicks effect, L_{ee} is the usual electrical conductivity, and L_{qe} is Thomsons' thermoelectric effect. As we shall see both of these cases play a very important roll in the physics of a ionized plasma.

The fourth assumption arises in some way from the structure of Eq. (1.5). Indeed viewing L_{ik} as the element of a matrix $\overleftrightarrow{L}$ it turns out that if it is diagonal, the constitutive equations lead to a complete set of "hydrodynamic equations", but if $\overleftrightarrow{L}$ is not diagonal, then more information is required for such purpose. This condition, buried in the very old approach of Lord Kelvin to the study of the thermoelectric effect was later identified by L. Onsager in 1931 in his study on chemical reactions and proved by the same author twenty years later for an arbitrary system. The condition is that $\overleftrightarrow{L}$ must be a symmetrical matrix whenever microscopic reversibility holds true. This requirement reflects the invariance of the microscopic equations of motion of the particle composing the system under time reversal. Thus if $\overleftrightarrow{L}^{\dagger}$ is the transposed of $\overleftrightarrow{L}$ then,

$$\overleftrightarrow{L} = \overleftrightarrow{L}^{\dagger} \tag{1.11}$$

is Onsagers' reciprocity theorem (ORT). In words, it is the ultimate way in which microscopic reversibility is exhibited on a macroscopic level. No thermodynamical theory based on Eq. (1.5) may be considered complete if Eq. (1.11) is not fulfilled.

Regretfully the literature is plagued with results which arise from misconceptions about the nature of (1.11). One thing is to obtain transport coefficients which are consistent with Eq. (1.11) and another thing is to be able to rigorously prove Eq. (1.11) from the basic kinetic equation from which $\overleftrightarrow{L}$ is calculated.

This is not the place to enter into a profound discussion on the nature of ORT but it is worth mentioning one example. If one computes $\overleftrightarrow{L}$ from a Boltzmann kinetic equation for a multicomponent mixture of dilute inert gases, the proof that Eq. (1.11) holds true has been a headache for years. Indeed, this equation may be shown to hold true if one selects the chemical potential as the thermodynamic force in Eqs. (1.9) provided the system is isothermic [3].

However if a temperature gradient is present this is no longer correct and Eq. (1.11) can be shown to hold if, and only if, the thermodynamic forces are represented by the diffusive force, to be thoroughly discussed in this book, and not by chemical potentials [9]. When, like in plasmas, magnetic fields are present the phenomenological form of Eq. (1.11) may be readily inferred but its microscopic derivation is far from trivial. In fact, when using the Landau-Fokker Planck version of Boltzmann' equation, the proof is impossible since the approximations introduced into the scheme erases the source of microscopic reversibility from the resulting kinetic equation. Thus, Eq. (1.11) may be checked to hold but not derived from the kinetic equation itself [10]. Consistency with Onsager's reciprocity theorem is often claimed but, we repeat, an air tight proof of its validity may not be so easy to obtain.

These general ideas provide the necessary background to understand what we mean by placing the derivation of the transport equations of dilute inert plasma within the framework of CIT. For more details we recommend the reader to seek more information in the vast literature on the subject.

Bibliography

[1] S. R. de Groot; *Thermodynamics of Irreversible Processes*; North-Holland Publ. Co., Amsterdam (1952).

[2] J. Meixner and H. Reik; *Thermodynamik der Irreversiblen Prozeses*, in Handbuch der Physik; S. Flügge, ed. Springer-Verlag, Berlin (1959), vol. 3.

[3] S. R. de Groot and P. Mazur; *Non-equilibrium Thermodynamics*; Dover Publications Inc., Mineola, N. Y. (1984).

[4] R. Hasse; *Thermodynamics of Irreversible Processes*; Addison-Wesley Publ. Co., Reading, Mass (1971).

[5] I. Prigogine; *Thermodynamics of Irreversible Processes*; Wiley-Interscience, New York (1967), 3rd ed.

[6] L. García-Colín and P. Goldstein; *La Física de Procesos Irreversibles*; El Colegio Nacional, México D. F. (2003), Vols. 1 and 2 (in Spanish)

[7] L. García-Colín and F. J. Uribe; *J. Non Equilib. Thermodyn.* **16**, 89 (1991).

[8] C. L. Fefferman, *Existence and Smoothness of the Navier-Stokes equations http://www.claymath.org/prizeproblems/navierstokes.htm*, 89 (1991).

[9] P. Goldstein and L. García-Colín; *J. Non Equilib. Thermodyn.* **30**, 173 (2005).

[10] R. Balescu; *Transport Processes in Plasmas*, North-Holland Publ. Co., Amsterdam (1988), Vol. 1: Classical Transport.

Chapter 2

The Problem

The system we wish to study is a binary mixture of non reactive dilute, electrically charged system of particles. Their masses will be labelled m_a and m_b with charges e_a and e_b where $e_a = -e_b = e$. The ions could have a positive charge Ze but we shall keep $Z = 1$ for simplicity. The number densities of the species are n_a and n_b where $n_a + n_b = n$ so that the total mass density ρ is given by

$$\rho = \rho_a + \rho_b = m_a n_a + m_b n_b \tag{2.1}$$

Following the standard notation of the kinetic theory of gases, the single particle distribution functions for each species is denoted by $f_i\left(\vec{r}, \vec{v_i}, t\right)$ where $\vec{v_i}$ is the velocity of the particle of species i, $i = a, b$. If we now assume that in general the system is acted upon by an electric field $\vec{E}$ measured in volts m^{-1} and a magnetic induction $\vec{B}$ in teslas, the Boltzmann equation determining the time evolution of the distribution function f_i is given by

$$\frac{\partial f_i}{\partial t} + \vec{v_i} \cdot \frac{\partial f_i}{\partial \vec{r}} + \frac{1}{m_i}\left(\vec{F_i} + e_i \vec{v_i} \times \vec{B}\right) \cdot \frac{\partial f_i}{\partial \vec{v_i}} = \sum_{i,j=a}^{b} J\left(f_i f_j\right) \tag{2.2}$$

Here,

$$\vec{F_i} = \vec{F_i^{(e)}} + e_i \vec{E} \quad \text{for } i, j = a, b \tag{2.3}$$

$\vec{F_i^{(e)}}$ denoting an external conservative force, the electric and magnetic fields, $\vec{E}$ and $\vec{B}$ respectively, are the self consistent fields generated by the plasma as determined from Maxwell's equations, and

$$J\left(f_i f_j\right) = \int \cdots \int \left\{ f(\vec{v_i'}) f(\vec{v_j'}) - f\left(\vec{v_i}\right) f\left(\vec{v_j}\right) \right\}$$
$$\sigma\left(\vec{v_i}\vec{v_j} \to \vec{v_i'}\vec{v_j'}\right) g_{ij} d\vec{v_j} d\vec{v_i'} d\vec{v_j'} \tag{2.4}$$

L.S. García-Colín, L. Dagdug, *The Kinetic Theory of a Dilute Ionized Plasma*,
Springer Series on Atomic, Optical, and Plasma Physics 53

In Eq. (2.4) we recall the reader that the $\vec{r}, t$ dependence of the f_i's has been omitted. The primes denote the values of v_i after the binary collision takes place, $\sigma\left(\vec{v_i}\vec{v_j} \to \vec{v_i'}\vec{v_j'}\right) d\vec{v_i'}d\vec{v_j'}$ is the cross section, namely, the number of molecules per unit time of species i colliding with a molecule of species j such that after the collision the molecules have velocities $\vec{v_i'}$ in the range $d\vec{v_i'}$ and $\vec{v_j'}$ in the range $d\vec{v_j'}$; $g_{ij} \equiv |\vec{v_i} - \vec{v_j}| = |\vec{v_i'} - \vec{v_j'}|$. For collisions between molecules of the same species $\vec{v_i} \to \vec{v}$ and $\vec{v_j} \to \vec{v_1}$ to distinguish the two velocities. A caution note has to be mentioned with respect of Eq. (2.2). The magnetic induction $\vec{B}$ is taken to be the average magnetic field, determined from Maxwell's equations where the current density will depend on the distribution functions f_i. In fact one should write $\vec{B} = \vec{B}_{av} + \vec{B}_e$ where $\vec{B}_e$ is an external field which may or may not be taken as a constant field.[1] We also recall the reader that the cross section σ satisfies the principle of microscopic reversibility, namely, it is invariant upon spatial and temporal reflections, so that,

$$\sigma\left(\vec{v_i}\vec{v_j} \to \vec{v_i'}\vec{v_j'}\right) = \sigma\left(\vec{v_i'}\vec{v_j'} \to \vec{v_i}\vec{v_j}\right) \quad \text{for} \quad i, j = a, b \tag{2.5}$$

thus guaranteeing the existence of inverse collisions.

As well known in kinetic theory, two general results may be derived from Eq. (2.4) regardless of the specific form of the cross section that is, without specifying the details of the interaction potential between the particles. Such results are the conservation equations and the H theorem. In our case this will require particular care since collisions do not exist for Coulomb interactions which as well known is a long range repulsive potential. Advancing the fact that this will be appropriately taken care of using the Debye-Hückel approximation we assume that σ is well defined and finite. We proceed to discuss the first of two general results namely, the conservation equations. Section 2.2 will be devoted to the H-theorem.

2.1 Conservation Equations

As usual, we define the local particle densities as,

$$n_i(\vec{r}, t) = \int f_i(\vec{r}, \vec{v_i}, t) d\vec{v_i} \tag{2.6}$$

[1] For a thorough discussion of this question see [7].

and denote by $\psi_i(\vec{r}, \vec{v_i}, t)$ any dynamical variable whose local value is given by

$$\langle\psi_i\rangle \equiv \psi_i(\vec{r}, t) = \frac{1}{n_i}\int \psi_i(\vec{r}, \vec{v_i}, t) f_i(\vec{r}, \vec{v_i}, t) d\vec{v_i} \tag{2.7}$$

Moreover, we define the thermal or chaotic velocity $\vec{c_i}$ as

$$\vec{c_i} = \vec{v_i} - \vec{u}(\vec{r}, t) \tag{2.8}$$

and $\vec{u}$ is the barycentric velocity given by

$$\rho\vec{u}(\vec{r}, t) = \sum_i \rho_i \vec{u_i}(\vec{r}, t) \tag{2.9}$$

where,

$$\vec{u_i}(\vec{r}, t) = \frac{1}{n_i}\int f_i(\vec{r}, \vec{v_i}, t)\vec{v_i} d\vec{v_i} \tag{2.10}$$

is the local hydrodynamic velocity for species i. Notice here that contrary to what occurs in the case of a single species, $\langle\vec{c_i}\rangle \neq 0$ whereas

$$m_a n_a \langle\vec{c_a}\rangle + m_b n_b \langle\vec{c_b}\rangle = \rho_a \vec{u_a} + \rho_b \vec{u_b} - \rho\vec{u} = 0$$

or

$$\sum_i \rho_i \langle\vec{c_i}\rangle = 0 \tag{2.11}$$

This expression is important because the mass diffusion flux of the ith species is defined as

$$\vec{J_i} = m_i \int \vec{c_i} f_i(\vec{r}, \vec{v_i}, t) d\vec{v_i} = m_i n_i \langle\vec{c_i}\rangle \tag{2.12}$$

so that by Eq. (2.11)

$$\sum_i \vec{J_i} = 0 \tag{2.13}$$

or $\vec{J_a} = -\vec{J_b}$.

With these definitions, the flow of charge is readily expressed in a convenient way. In fact, the numerical charge density Q is defined as

$$Q = n_a e_a + n_b e_b = (n_a - n_b)e \tag{2.14a}$$

and the charge current

$$\vec{J_T} = \sum_i n_i e_i \langle\vec{v_i}\rangle \tag{2.14b}$$

which, with the aid of Eq. (2.8) reads $\vec{J}_T = Q\vec{u} + \vec{J}_c$, $\vec{J}_c$ the conduction current being given by

$$\vec{J}_c = \sum_i n_i e_i \langle \vec{c}_i \rangle \tag{2.14c}$$

which in turn can be written with the aid of Eqs. (2.12) and (2.13) as

$$\vec{J}_c = \frac{m_a + m_b}{m_a m_b} e\vec{J}_a \tag{2.15}$$

a result often ignored by authors of this subject.

Returning to our quest, we now derive the equivalent of Maxwell-Enskog's transport equation by taking $\psi_i = (m_i, m_i\vec{v}_i$ and $\frac{1}{2}m_i\vec{v}_i^2)$. We first notice that from Eq. (2.4)

$$\sum_{i,j=a}^{b} \int \psi_i J(f_i f_j) d\vec{v}_i = 0 \tag{2.16}$$

a result which follows from the standard transformation of the collision kernels using Eq. (2.5) and the fact that i and j are dummy indices in Eq. (2.16).

So let $\psi_i = m_i$. Multiplying (2.2) by m_i and integrating over $d\vec{v}_i$ using (2.16) one gets

$$\frac{\partial \rho_i}{\partial t} + \text{div}\ (\rho_i \vec{u}_i) = \int (\vec{v}_i \times \vec{B}) \cdot \frac{\partial f_i}{\partial \vec{v}_i} d\vec{v}_i$$

In the right hand term, for any component $\dfrac{\partial f_i}{\partial \vec{v}_i}$ the cross product $(\vec{v}_i \times \vec{B})$ does not contain such component so that the integration by parts yields zero whence

$$\frac{\partial \rho_i}{\partial t} + \text{div}\ (\rho_i \vec{u}_i) = 0 \tag{2.17a}$$

and summation over i yields

$$\frac{\partial \rho}{\partial t} + \text{div}\ (\rho \vec{u}) = 0 \tag{2.17b}$$

Using Eqs. (2.8) and (2.12) in Eq. (2.17a) we may also write that

$$\frac{\partial \rho_i}{\partial t} + \text{div}\ (\rho_i \vec{u}) = -\text{div}\ \vec{J}_i \tag{2.17c}$$

Eqs. (2.17a)-(2.17c) are thus the several alternative expressions for mass conservation.

Take now $\psi_i = m_i\vec{v}_i = m_i(\vec{c}_i + \vec{u})$. Multiply Eq. (2.2) by it using Eq. (2.11), after integrating a couple of terms by parts and summing over i, one readily gets that

$$\frac{\partial}{\partial t}(\rho\vec{u}) + \text{div}\,(\overleftrightarrow{\tau}^k + \rho\vec{u}\vec{u}) = \sum_i n_i\vec{F}_i - \sum_i e_i \int \vec{v}_i(\vec{v}_i \times \vec{B}) \cdot \frac{\partial f_i}{\partial \vec{v}_i} d\vec{v}_i$$

where the kinetic part of the stress tensor $\overleftrightarrow{\tau}^k$ is defined as

$$\overleftrightarrow{\tau}^k = \sum_{i=a}^{b} m_i \int f_i\vec{c}_i\vec{c}_i d\vec{v}_i \tag{2.18}$$

Integration by parts of the last term reduces to $\sum_i e_i n_i\langle\vec{v}_i\rangle \times \vec{B}$ so that we reach the result that

$$\frac{\partial}{\partial t}(\rho\vec{u}) + \text{div}\,(\overleftrightarrow{\tau}^k + \rho\vec{u}\vec{u}) = \sum_i n_i\vec{F}_i + (\vec{J}_T \times \vec{B}) \tag{2.19}$$

the conservation equation for momentum. If the external force is zero using the definition of $\vec{J}_T$ we readily find that

$$\frac{\partial}{\partial t}(\rho\vec{u}) + \text{div}\,(\overleftrightarrow{\tau}^k + \rho\vec{u}\vec{u}) = Q(\vec{E} + (\vec{u} \times \vec{B})) + \vec{J}_c \times \vec{B} \tag{2.20}$$

Here $\vec{E}' = \vec{E} + \vec{u} \times \vec{B}$ can be interpreted as the effective electric field as viewed by an observer moving in the mixture with the barycentric velocity $\vec{u}$. Also, it should be pointed out that often Eqs. (2.17a)-(2.17c) and (2.20) are referred to as the equations of magnetohydrodynamics for isothermal fluids in the absence of external fields $\vec{F}^e = 0$.

We finally take $\psi_i = \frac{1}{2}m_i v_i^2$ and repeat the procedure as in the previous case. After summation over i and use of Eq. (2.11) we get that,

$$\frac{1}{2}\frac{\partial}{\partial t}(\rho u^2) + \frac{\partial}{\partial t}\sum_i \frac{1}{2}m_i \int f_i d\vec{v}_i c_i^2 + \sum_i \frac{1}{2}m_i \int \vec{v}_i \cdot \frac{\partial f_i}{\partial \vec{r}} v_i^2 d\vec{v}_i +$$
$$\sum_i \frac{1}{2} \int \vec{F}_i \cdot v_i^2 \frac{\partial f_i}{\partial \vec{v}_i} d\vec{v}_i + \frac{1}{2}\sum_i e_i \int (\vec{v}_i \times \vec{B}) \cdot \frac{\partial f_i}{\partial \vec{v}_i} v_i^2 d\vec{v}_i = 0 \tag{2.21}$$

We define the internal energy density of the mixture as,

$$\rho e(\vec{r}, t) = \sum_i \frac{1}{2}\rho_i\langle c_i^2\rangle \tag{2.22}$$

In the third term we set $\vec{v_i} = \vec{c_i} + \vec{u}$, expand, use Eq. (2.11) and find that it reduces to

$$\mathrm{div}(\vec{J_q} + \vec{u} \cdot \overleftrightarrow{\tau}^k + \rho e \vec{u} + \frac{1}{2}\rho \vec{u} u^2)$$

where

$$\vec{J_q} = \sum_i \frac{1}{2}\rho_i \langle \vec{c_i} c_i^2 \rangle \tag{2.23}$$

is the total heat flux in the mixture. After a first integration by parts, use of (2.14c), the definition of $\vec{J_T}$ and assuming $\vec{F^e} = \vec{0}$, the fourth term simply reduces to $-\vec{J_T} \cdot \vec{E}$. Finally integration by parts clearly shows that the last term vanishes, so that collecting all terms we find that,

$$\frac{1}{2}\frac{\partial}{\partial t}(\rho u^2) + \frac{\partial \rho e}{\partial t} + \mathrm{div}\ (\vec{J_q} + \vec{u} \cdot \overleftrightarrow{\tau}^k + \rho \vec{u} e + \frac{1}{2}\rho \vec{u} u^2) - \vec{J_T} \cdot \vec{E} = 0$$

Using Eq. (2.20) and following the standard steps to combine the first three terms in this equation we are finally lead to the balance equation for the internal energy namely,

$$\rho \frac{d}{dt} e + \mathrm{div}\ \vec{J_q} + \overleftrightarrow{\tau}^k : \mathrm{grad}\ \vec{u} - \vec{J_c} \cdot \vec{E'} = 0 \tag{2.24}$$

where as introduced above,

$$\vec{E'} = \vec{E} + \vec{u} \times \vec{B}$$

Eqs. (2.17a)-(2.17c), (2.20) and (2.24) are the sought result for the conservation equations. Clearly the unknowns $\vec{J_i}$, $\overleftrightarrow{\tau}^k$ and $\vec{J_q}$ have to be determined by seeking solutions to Eq. (2.2), a task to be dealt with later.

2.2 The H Theorem and Local Equilibrium

Before discussing these important properties of the Boltzmann equation we need to specify clearly the domain of its applicability. In the absence of a magnetic field Eq. (2.2) is valid in the so called kinetic regime characterized by time $t \sim \tau$ the mean free time where $\tau \gg t_c$ the duration of a collision time. However, in the presence of a magnetic field we have two characteristic frequencies competing in the mixture, the collision frequency $\omega_c \sim 1/\tau$ and the Larmor frequencies $\omega_i = \frac{|e|B}{m_i}$. For electrons $\omega_e \sim 1.76 \times 10^{11} B$ whereas for ions $\omega_i = \omega_e \frac{m_e}{m_i}$. If the field is weak enough $\omega_i \tau$ is of the order of 1 for both cases implying that the field does not interfere in the collisional regime of the mixture. We shall limit ourselves to this case. When $\omega_i \tau \gg 1$ radical modifications have to be made to the whole approach to the problem and we shall not discuss it here at all (see however Ref. [7]). Once this is clarified we proceed with our discussion. If we multiply Eq. (2.2) by $\ln f_i$ integrate over $d\vec{v_i}$ and sum over i the left hand side vanishes since the only extra term, $\int (\vec{v_i} \times \vec{B}) \cdot \frac{\partial f_i}{\partial \vec{v_i}} \ln f_i d\vec{v_i}$ vanishes after integration by parts. Therefore, using the same procedure for the right hand side as in the single component case remembering Eq. (2.2) and Klein's inequality one obtains that for

$$H \equiv \sum_i \int f_i \ln f_i d\vec{v_i} \quad , \tag{2.25}$$

$$\frac{\partial H(\vec{r}, t)}{\partial t} \leq 0 \tag{2.26}$$

for all binary collisions and their corresponding inverses $(i, j \rightleftharpoons i', j')$. Remember that in Eq. (2.5), $H \equiv H(\vec{r}, t)$ is still function of $\vec{r}$ and t. So the irreversibility criteria imposed by Eq. (2.26) is still valid in the weak field approximation and moreover, the quantity usually associated with the entropy production $\sigma(\vec{r}, t)$ is always positive definite for all exact solutions to Eq. (2.4)

$$\sigma = -k \sum_{i,j} \int \ln f_i J(f_i f_j) d\vec{v_i} \tag{2.27}$$

This result will be used later on. We also notice that the solution to the homogenous Boltzmann equation, namely,

$$J(f_i^{(0)} f_i^{(0)}) + J(f_i^{(0)} f_j^{(0)}) = 0 \quad \text{for} \quad i, j = a, b$$

is a local Maxwellian distribution function. This arises from the well known argument stating that $\frac{\partial H}{\partial t} = 0$ for every binary collision. By the standard argument of kinetic theory this implies that

$$f_i^{(0)} = n_i(\vec{r}, t) \left(\frac{m_i}{2\pi k T(\vec{r}, t)} \right)^{\frac{3}{2}} e^{-\frac{m(\vec{v_i} - \vec{u}(\vec{r}, t))^2}{2kT(\vec{r}, t)}} \tag{2.28}$$

provided we define

$$n_i(\vec{r}, t) = \int f_i^{(0)} d\vec{v_i} \tag{2.29a}$$

$$\rho\vec{u} = \sum_i \rho_i \int f_i^{(0)} \vec{v}_i d\vec{v}_i \tag{2.29b}$$

$$\rho e(\vec{r},t) = \frac{3}{2} nkT = \sum_i \frac{1}{2} \rho_i \langle c_i^2 \rangle \tag{2.29c}$$

Nevertheless Eq. (2.28) is still not a solution to the full Boltzmann equation since it is necessary that

$$\left(\frac{\partial}{\partial t} + \vec{v}_i \cdot \frac{\partial}{\partial \vec{r}} + \frac{\vec{F}_i}{m_i} \cdot \frac{\partial}{\partial \vec{v}_i} + \frac{e_i}{m_i} (\vec{v}_i \times \vec{B}) \cdot \frac{\partial}{\partial \vec{v}_i} \right) \ln f_i^{(0)} = 0 \tag{2.30}$$

is satisfied for $i = a, b$. The procedure is, once more, the standard one [1]-[2]. We write

$$\ln f_i^{(0)} = \nu(\vec{r},t) + \vec{k}(\vec{r},t) \cdot \vec{v}_i - h(\vec{r},t) v_i^2 \tag{2.31}$$

where $\nu = \ln A - \frac{m_i \beta}{2} u^2$, $\vec{k} = \beta m_i \vec{u}$; $h = \frac{m_i \beta}{2}$; $A = n_i \left(\frac{m_i \beta}{2\pi} \right)^{\frac{3}{2}}$ with $\beta = (kT)^{-1}$.

Substitution of (2.31) into (2.30) and noticing that $(\vec{v}_i \times \vec{B}) \cdot \vec{v}_i = 0$ we get that,

$$\frac{\partial \nu}{\partial t} - v_i^2 \vec{v}_i \cdot \frac{\partial h}{\partial \vec{r}} + \left(-v_i^2 \frac{\partial h}{\partial t} + \vec{v}_i \cdot \left(\vec{v}_i \cdot \frac{\partial \vec{k}}{\partial \vec{r}} \right) \right) +$$

$$\vec{v}_i \cdot \left(\frac{\partial \vec{k}}{\partial t} + \frac{\partial \nu}{\partial \vec{r}} - \frac{2h}{m_i} \vec{F}_i \right) + \vec{k} \cdot (\vec{v}_i \times \vec{B}) \frac{e_i}{m_i} + \frac{\vec{F}_i}{m_i} \cdot \vec{k} = 0$$

which must hold for all values of $\vec{v}$. The coefficients of order v_i^3 and v_i^2 do not depend on $\vec{B}$ so by the standard procedure $h = h(t)$ and $\vec{k} = \vec{r} \frac{\partial h}{\partial t} + \vec{r} \times \vec{\Omega}(t) + \vec{k}_0(t)$. For conservative forces (including $\vec{F} = -e \operatorname{grad} \phi$) the linear coefficient in $\vec{v}_i$ yields

$$\frac{\partial \vec{k}}{\partial t} + \operatorname{grad} \left(\nu + \frac{2he_i}{m_i} \phi_i \right) - \frac{e_i}{m_i} \vec{k} \times \vec{B} = 0$$

and ϕ_i is the electrical potential. Scalar multiplication by $\vec{B}$, yields in turn that

$$\vec{B} \cdot \left(\frac{\partial \vec{k}}{\partial t} + \operatorname{grad} \left(\nu + \frac{2he_i}{m_i} \phi_i \right) \right) = 0$$

which for $\vec{B} \neq \vec{0}$ and ignoring the possible but unlikely occurrence that $\vec{B}$ is perpendicular to the term in parenthesis,

$$\frac{\partial \vec{k}}{\partial t} + \text{grad}\left(\nu + \frac{2he_i}{m_i}\phi_i\right) = 0$$

This implies $\frac{\partial}{\partial t}\text{rot}\vec{k} = \vec{0}$ or $\vec{\Omega}$ is a constant vector and, once more by the argument for a one component system, and non-pathological external forces,

$$f_i^{eq} = n_i\left(\frac{m_i}{2\pi kT}\right)^{\frac{3}{2}} \exp\left\{-\beta\left(\frac{m_i v_i^2}{2} + \phi_i(\vec{r})\right)\right\} \quad \text{for } i = a, b \tag{2.32}$$

where the potential energy is $\phi_i = \phi_{ext} + e_i\phi$. Thus equilibrium is achieved and characterized by the Maxwell distribution function Eq. (2.32).

Bibliography

[1] S. Chapman and T. G. Cowling; *The Mathematical Theory of Non-Uniform Gases*; Cambridge University Press, Cambridge (1970), 3rd ed.

[2] G. W. Ford and G. E. Uhlenbeck; *Lectures in Statistical Mechanics*; American Mathematical Society, Providence, R. I. (1963).

[3] J. R. Dorfman and H. van Beijren; *The Kinetic Theory of Gases*, in Statistical Mechanics, Pt. B.; Bruce J. Berne, ed. Plenum Press, New York (1977).

[4] J. H. Ferziger and H. G. Kaper; *Mathematical Theory of Transport Processes in Gases*; North-Holland Publ. Co., Amsterdam (1972).

[5] C. Cercignani; *The Boltzmann Equation and its Applications*; Springer-Verlag, New York (1988).

[6] L. García-Colín and P. Goldstein; *La Física de los Procesos Irreversibles*; Vols. 1 and 2, El Colegio Nacional, Mexico D. F. (2003) (in Spanish).

[7] R. Balescu; *Transport Processes in Plasma Vol. I, Classical Transport*; North-Holland Publ. Co., Amsterdam (1988).

[8] W. Marshall; *The Kinetic Theory of an Ionized Gas*; U.K.A.E.A. Research Group, Atomic Energy Research Establishment. Harwell U.K. parts I, II, and III (1960).

Chapter 3

Solution of the Boltzmann Equation

In this section we want to discuss the solution to Eq. (2.2). The first problem we encounter concerns the correct interpretation of the drift term $\vec{v_i} \times \vec{B}$ where $\vec{v_i}$ is the velocity of a particle of species i. Since $\vec{v_i} = \vec{c_i} + \vec{u}$ and $|\vec{c_i}| >> |\vec{u}|$ being the velocity associated with the thermal agitation of the molecules we follow Chapman and Cowling's suggestion in keeping $\vec{u} \times \vec{B}$ in the drift term and bring $\vec{c_i} \times \vec{B}$ to the collisional contribution. Using the definition of the "effective" electric field $\vec{E}'$, Eq. (2.2) may be rewritten as follows,

$$\frac{\partial f_i}{\partial t} + \vec{v_i} \cdot \frac{\partial f_i}{\partial \vec{r}} + \frac{1}{m_i}\left(\vec{F_i} + e_i \vec{E}'\right) \cdot \frac{\partial f_i}{\partial \vec{v_i}} = -\frac{e_i}{m_i}(\vec{c_i} \times \vec{B}) \cdot \frac{\partial f_i}{\partial \vec{v_i}} + \sum J\left(f_i f_j\right)$$
$$\text{for } i, j = a, b \tag{3.1}$$

Notice that when $\vec{B} = \vec{0}$, $\vec{F_i} + e_i \vec{E}$ is a conservative force whence the solution to Eq. (3.1) is trivial extension of the solution obtained for an inert mixture in the presence of a conservative force. What will complicate the solution to Eq. (3.1) is the presence of the first term in the r.h.s. As usual, we now assume that in the weak field approximation the distribution functions $f_i(\vec{r}, \vec{v_i}, t)$ can be taken as functionals of the locally conserved variables namely $f_i(\vec{r}, \vec{v_i} | n_i(\vec{r}, t), u(\vec{r}, t), e(\vec{r}, t))$[1] and further, they may be expanded in power series of Knudsen's parameter ϵ around the local equilibrium distribution function $f_i^{(0)}$ defined in Eq. (2.28). Omitting unnecessary arguments,

$$f_i = f_i^{(0)}(1 + \epsilon \varphi_i^{(1)} + \epsilon^2 \varphi_i^{(2)} + \cdots) \quad \text{for} \quad i = a, b \tag{3.2}$$

[1]Notice that $\vec{u_i}$ is not taken as a local variable.

L.S. García-Colín, L. Dagdug, *The Kinetic Theory of a Dilute Ionized Plasma*, Springer Series on Atomic, Optical, and Plasma Physics 53

the well known Hilbert-Chapman-Enskog approximation. We remind the reader that the parameter ϵ is a measure of the unitary change of a local variable in a mean free path. Notice that since

$$\frac{\partial f_i^{(0)}}{\partial \vec{v}_i} = -f_i^{(0)} \frac{m_i}{kT} \vec{c}_i$$

and that $(\vec{c}_i \times \vec{B}) \cdot \vec{c}_i = \vec{0}$, $f_i^{(0)}$ is still a solution to the homogeneous part of Eq. (3.1) as stated in page 19. Substitution of Eq. (3.2) in Eq. (3.1), collecting terms of order ϵ and for clarity taking $i = a$, Eq. (3.1) reduces to

$$\frac{\partial f_a^{(0)}}{\partial t} + \vec{v}_a \cdot \frac{\partial f_a^{(0)}}{\partial \vec{r}} + \frac{1}{m_a} \left(\vec{F}_a + e_a \vec{E}' \right) \cdot \frac{\partial f_a^{(0)}}{\partial \vec{v}_a} =$$

$$-\frac{e_a}{m_a} (\vec{c}_a \times \vec{B}) \cdot \frac{\partial}{\partial \vec{v}_a} \left(f_a^{(0)} \varphi_a^{(1)} \right) + f_a^{(0)} \left\{ C(\varphi_a^{(1)}) + C(\varphi_a^{(1)} + \varphi_b^{(1)}) \right\} \tag{3.3}$$

where $C(\varphi_a^{(1)})$ and $C(\varphi_a^{(1)} + \varphi_b^{(1)})$ are the linearized collision kernels for collisions among particles of species a and species a, b, respectively. Their explicit form will be written later.[2] Clearly there is an identical equation for species b simply obtained by exchanging the indices a and b in (3.3). As the next step we must evaluate explicitly the left hand side of Eq. (3.3) since by the functional assumption,

$$\frac{\partial f_a^{(0)}}{\partial t} = \frac{\partial f_a^{(0)}}{\partial n_i} \frac{\partial n_i}{\partial t} + \frac{\partial f_a^{(0)}}{\partial \vec{u}} \cdot \frac{\partial \vec{u}}{\partial t} + \frac{\partial f_a^{(0)}}{\partial T} \frac{\partial T}{\partial t}$$

where the local temperature $T(\vec{r}, t)$ is introduced as usual through the equation

$$\rho e(\vec{r}, t) = \frac{3}{2} nkT \quad , \quad n = n_a + n_b \tag{3.4}$$

The time derivatives of the local variables are to be computed to first order in the gradients namely, through the Euler equations. Clearly, a similar expression holds for $\frac{\partial f_a^{(0)}}{\partial \vec{r}}$ which is unnecessary to write down explicitly. Also,

$$\frac{\partial f_a^{(0)}}{\partial n_a} = \frac{f_a^{(0)}}{n_a} \; ; \; \frac{\partial f_a^{(0)}}{\partial \vec{u}} = \frac{m_a}{kT} \vec{c}_a f_a^{(0)} \; ; \; \frac{\partial f_a^{(0)}}{\partial T} = \frac{f_a^{(0)}}{T} \left(\frac{m_a c_a^2}{2kT} - \frac{3}{2} \right) \tag{3.5}$$

[2] See Eqs. (31), (32) in Ref. [1].

Euler's equations are readily obtained as follows. Firstly, from Eq. (2.17c) (the local variable is $\vec{u}$!) noticing that $J_i^{\vec{e}q} = \vec{0}$ since $\langle \vec{c_i} \rangle_{eq} = \vec{0}$ and div $\langle \vec{c_i} \rangle \sim \epsilon^2$ we have

$$\left(\frac{\partial n_a}{\partial t}\right)_0 = -\text{div } n_a \vec{u} \tag{3.6}$$

where subscript 0 is to emphasize the lowest (first) order in the gradients term. In Eq. (2.19) since $\frac{\partial(\rho \vec{u})}{\partial t} + \vec{u} \text{ div } \vec{u} = \rho \frac{d\vec{u}}{dt}$ we do not ignore the non-linear term $\vec{u} \cdot \text{grad } \vec{u}$; also $\overleftrightarrow{\tau} = p\mathbb{I}$ and $\vec{J_c} \sim \epsilon$ since $\vec{J_a}$ is a first order in ϵ term. Whence,

$$\rho \left(\frac{\partial \vec{u}}{\partial t}\right)_0 = -\rho \vec{u} \cdot \text{grad } \vec{u} - \text{grad } p + Q\vec{E}' + \vec{J_c} \times \vec{B} + \sum_i n_i \vec{F_i} \tag{3.7}$$

Finally, since from Eq. (2.17b) $\rho \frac{d}{dt} e = \frac{de}{dt} + \rho e \text{ div } \vec{u}$ using Eqs. (3.4), and (3.6) it is readily seen that $\frac{de}{dt} + \rho e \text{ div } \vec{u} = \frac{3}{2} nk \frac{dT}{dt}$ and since $\vec{J_c} \cdot \vec{E}' \sim \epsilon^2$ because $\vec{E}'$ is a first order in the gradients term, div $\vec{J_q} \sim \epsilon^2$, one finally arrives at the result,

$$\left(\frac{\partial T}{\partial t}\right)_0 = -\left(\vec{u} \cdot \text{grad } T + \frac{2p}{3nk} \text{ div } \vec{u}\right) \tag{3.8}$$

Eqs. (3.6)-(3.8) are the Euler equations of magnetohydrodynamics.

Using Eqs. (3.6)-(3.8), the corresponding equation for $\frac{\partial f_a^{(0)}}{\partial r}$ and writing the explicit form for $\vec{J_c}$ in Eq. (3.7), after some tedious algebra one arrives at the following expression namely,

$$f_a^{(0)} \left\{ \frac{m_a}{kT} \overleftrightarrow{c_a{}^0 c_a} : \text{grad } \vec{u} + \left(\frac{m_a c_a^2}{2kT} - \frac{5}{2}\right) \text{grad } \ln T \cdot \vec{c_a} + \frac{n_a}{n} \vec{c_a} \cdot \vec{d_{ab}} \right\} =$$

$$-f_a^{(0)} \frac{e_a}{m_a} \left(\vec{c_a} \times \vec{B}\right) \cdot \frac{\partial \varphi_a^{(1)}}{\partial \vec{v}} - \frac{m_a}{\rho kT} f_a^{(0)} \left\{ \sum_{j=a}^{b} e_j \left[\int d\vec{c_j} f_j^{(0)} \varphi_j^{(1)} \vec{c_j}\right] \times \vec{B} \right\} \cdot \vec{c_a} +$$

$$f_a^{(0)} \left\{ C\left(\varphi_a^{(1)}\right) + C\left(\varphi_a^{(1)} + \varphi_b^{(1)}\right) \right\} \tag{3.9}$$

and an identical equation for species b. In Eq. (3.9) $\overleftrightarrow{A}^0$ denotes the symmetric traceless part of any tensor A. Here,

$$\vec{d_{ab}} = -\vec{d_{ba}} = \text{grad}\ \frac{n_a}{n} + \frac{n_a n_b (m_b - m_a)}{n\rho}\frac{\text{grad}\ p}{p}$$

$$-\frac{\rho_a \rho_b}{p\rho}\left(\frac{\vec{F_a}}{m_a} - \frac{\vec{F_b}}{m_b}\right) - \frac{n_a n_b}{p\rho}(m_b e_a - m_a e_b)\cdot \vec{E'} \tag{3.10}$$

is the well known diffusive force accounting for all external forces and gradients in the system, others than grad$T$, which induces in the system the corresponding currents. In particular the $\vec{E}$ dependent term in $\vec{E'}$, is responsible for all the electrical effects in the mixture arising from the presence of the electric field. Notice further the fact that the solution to Eq. (3.9) is now severely complicated by the presence of the first two terms in its r.h.s. containing $\vec{B}$ and which really act as "collisional" contributions to $\varphi_i^{(1)}$.

The solution to Eq. (3.9) may be constructed by the standard procedure. We recall that when $\vec{B} = \vec{0}$, Eq. (3.9) has no unique solutions since the homogeneous equation has five particular solutions, the collisional invariants. Hence there are solutions composed by a solution to the inhomogeneous term plus an arbitrary linear combination of the solutions to the homogeneous equation. The coefficients appearing therein are determined through the subsidiary conditions namely,

$$\sum_i \int f_i^{(0)} \varphi_i^{(n)} \left\{ \begin{array}{c} m_i \\ m_i \vec{c_i} \\ \frac{1}{2} m_i c_i^2 \end{array} \right\} d\vec{c_i} = 0 \quad n \geq 1 \tag{3.11}$$

arising form the fact that we have arbitrarily chosen to determine the local variables only through $f_i^{(0)}$.

The argument is still valid for Eq. (3.9) except that we now notice that the homogeneous equation only has m_a and $\frac{1}{2} m_a c_a^2$ as particular solutions since, as the reader may readily verify the first two terms on the right hand side do not vanish for the collisional invariant $m_a \vec{c_a}$. Recall that $\frac{\partial \varphi_a}{\partial \vec{v_a}} = \frac{\partial \varphi_a}{\partial \vec{c_a}}$ since $\vec{u}$ is constant in $\vec{v}$-space, so that,

$$\varphi_i^{(1)} = \overleftrightarrow{\mathbb{B}_i}(\vec{c_i}, T, B \cdots) : \text{grad}\ \vec{u} + \vec{\mathbb{A}}_i(\vec{c_i}, T, B \cdots) \cdot \text{grad}\ \ln T +$$
$$\vec{\mathbb{D}}_i(\vec{c_i}, T, B \cdots) \cdot \vec{d_{ij}} + \alpha_1 + \alpha_2 m_i c_i^2 \tag{3.12}$$

where $\overleftrightarrow{\mathbb{B}}$, $\vec{\mathbb{A}}$ and $\vec{\mathbb{D}}$ are to be determined. Since these coefficients are constructed from the vectors $\vec{c_i}$ and $\vec{B}$ as we shall see below the resulting integrals, when (3.12) is substituted into the first and third condition in (3.11), are odd in $\vec{c_i}$ and therefore vanish. Thus the contributions from $\alpha_1 + \alpha_2 m_i c_i^2$ can be shown to vanish following the same arguments as in the non magnetic case and proven that $\alpha_1 = \alpha_2 = 0$. Thus,

$$\varphi_i^{(1)} = \overleftrightarrow{\mathbb{B}_i} : \text{grad } \vec{u} + \vec{\mathbb{A}_i} \cdot \text{grad } \ln T + \vec{\mathbb{D}_i} \cdot d_{ij} \tag{3.13}$$

is the most general solution to (3.9) where the coefficients are the most general tensor $\overleftrightarrow{\mathbb{B}}$, and vectors $\vec{\mathbb{A}}$, $\vec{\mathbb{D}}$ which can be constructed from the only base vectors available $\vec{c_i}$ and the vector $\vec{B}$. These are (non-skew vectors)

$$\vec{c_i} \ , \ \vec{c_i} \times \vec{B} \ \text{ and } \ (\vec{c_i} \times \vec{B}) \times \vec{B} \tag{3.14}$$

since $[\left(\vec{c_i} \times \vec{B}\right) \times \vec{B}] \times \vec{B} = -B^2 \vec{c_i} \times \vec{B}$.

The term $\overleftrightarrow{\mathbb{B}}$: grad $\vec{u}$ where $\overleftrightarrow{\mathbb{B}}$ is now the most general tensor that may be constructed from the set (3.14) and gives rise to viscomagnetic effects will be ignored in the subsequent discussion. This tantamounts to taking a shear free mixture for which grad$\vec{u} = \vec{0}$.

For the remaining two terms we have that

$$\vec{\mathbb{A}_i} = \mathbb{A}_{\mathbb{I}}(c_i, B, \cdots)\vec{c_i} + \mathbb{A}_{\mathbb{II}}(c_i, B, \cdots)\vec{c_i} \times \vec{B} + \mathbb{A}_{\mathbb{III}}(c_i, B, \cdots)(\vec{c_i} \times \vec{B}) \times \vec{B} \tag{3.15a}$$

$$\vec{\mathbb{D}_i} = \mathbb{D}_{\mathbb{I}}(c_i, B, \cdots)\vec{c_i} + \mathbb{D}_{\mathbb{II}}(c_i, B, \cdots)\vec{c_i} \times \vec{B} + \mathbb{D}_{\mathbb{III}}(c_i, B, \cdots)(\vec{c_i} \times \vec{B}) \times \vec{B} \tag{3.15b}$$

for $i = a$, b. The quantities $\mathbb{A}_{\mathbb{I}}$, $\mathbb{D}_{\mathbb{I}}$, $\cdots$ etc. are functions of all the scalars that can be formed with these vectors namely $c_i = |\vec{c_i}|$, $B = |\vec{B}|$, $T(\vec{r_i}, t)$, $\cdots$ etc. so from here on we shall omit writing their arguments.

Even in this simplified case, what follows is much more cumbersome to deal with than in the $\vec{B} = \vec{0}$ case. When Eqs. (3.15a), (3.15b) are substituted back in to Eq. (3.13) and this in turn plugged into the expressions for $\vec{J_q}$ and $\vec{J_i}$ we will get all direct effects like heat conduction and mutual diffusion plus the electromagnetic ones arising from the force $\vec{E}'$, plus the cross effects such as the Dufour and Soret effects and others. The program is to see if one can construct a matrix $\overleftrightarrow{L}$ relating all of them and such that $\overleftrightarrow{L} = \overleftrightarrow{L}^\dagger$, the Onsager reciprocity relations. For the much simpler case when $\vec{B} = \vec{0}$ this is readily accomplished by a direct extension of our recent work in ordinary mixtures (Ref. [1]) and the reader is referred to the original source. When $\vec{B} \neq \vec{0}$ this is still a challenge and an attempt to accomplish it, will be done elsewhere.

We start by substituting Eq. (3.13) back into (3.9) and equating the coefficients of grad $\ln T$ and $\vec{d_{ab}}$, respectively. This leads to two integral equations for the vector functions $\vec{\mathbb{A}}_i$ and $\vec{\mathbb{D}}_i$ which read as:

$$f_a^{(0)}\left(\frac{m_a c_a^2}{2kT}-\frac{5}{2}\right)\vec{c_a}=-f_a^{(0)}\frac{e_a}{m_a}(\vec{c_a}\times\vec{B})\cdot\frac{\partial\vec{\mathbb{A}}_a}{\partial\vec{c_a}}$$

$$-f_a^{(0)}\frac{m_a}{\rho kT}\left\{\sum_j e_j\left[\int d\vec{c_j}f_j^{(0)}\vec{\mathbb{A}}_j\vec{c_j}\right]\times\vec{B}\right\}\cdot\vec{c_a}+f_a^{(0)}\left\{C(\vec{\mathbb{A}}_a)+C(\vec{\mathbb{A}}_a+\vec{\mathbb{A}}_b)\right\}$$

(3.16a)

and

$$\frac{n_a}{n}\vec{c_a}f_a^{(0)}=-f_a^{(0)}\frac{e_a}{m_a}(\vec{c_a}\times\vec{B})\cdot\frac{\partial\vec{\mathbb{D}}_a}{\partial\vec{c_a}}$$

$$-f_a^{(0)}\frac{m_a}{\rho kT}\left\{\sum_j e_j\left[\int d\vec{c_j}f_j^{(0)}\vec{\mathbb{D}}_j\vec{c_j}\right]\times\vec{B}\right\}\cdot\vec{c_a}+f_a^{(0)}\left\{C(\vec{\mathbb{D}}_a)+C(\vec{\mathbb{D}}_a+\vec{\mathbb{D}}_b)\right\}$$

(3.16b)

and two identical equations for species b. The vectors $\vec{\mathbb{A}}_a$ and $\vec{\mathbb{D}}_a$ are themselves given by Eqs. (3.15a) and (3.15b) so that still involved manipulations have to be carried out to simplify these equations. Due to the similarity of the terms involved in both equations, it is only necessary to give the details for a single case. Noticing that $(\vec{c_j}\times\vec{B})\times\vec{B}=\vec{B}(\vec{c_j}\cdot\vec{B})-B^2\vec{c_j}$ we rewrite Eq. (3.15a) as

$$\vec{\mathbb{A}}_a=(\mathbb{A}_a^{(1)}-B^2\mathbb{A}_a^{(3)})\vec{c}_a+\vec{c}_a\times\vec{B}\mathbb{A}_a^{(2)}+\vec{B}(\vec{c}_a\cdot\vec{B})\mathbb{A}_a^{(3)}$$

where $\mathbb{A}_a^{(1)}\equiv\mathbb{A}_{\mathrm{I}}$, etc. and $\mathbb{A}_a^{(1)}$, $\mathbb{A}_a^{(2)}$, $\mathbb{A}_a^{(3)}$ are scalar functions of $|\vec{c}_i|$, $|\vec{B}|$ and $|\vec{c}_i\cdot\vec{B}|$ although we shall arbitrarily neglect here the dependence on the latter term. Also, $\mathbb{A}_a^{(1)}-B^2\mathbb{A}_a^{(3)}$ is a coefficient that is a function of the scalars B^2, etc. so that we shall simply relabel it as $\mathbb{A}_a^{(1)}$. Thus

$$\vec{\mathbb{A}}_a=\mathbb{A}_a^{(1)}\vec{c}_a+(\vec{c}_a\times\vec{B})\mathbb{A}_a^{(2)}+\vec{B}(\vec{c}_a\cdot\vec{B})\mathbb{A}_a^{(3)} \qquad (3.15')$$

As shown in Appendix A when this function is substituted in (3.16a) the first two terms on the right side yield

$$-f_a^{(0)}\frac{e_a}{m_a}(\vec{c_a}\times\vec{B})\cdot\frac{\partial\vec{\mathbb{A}}_a}{\partial\vec{c_a}}=-f_a^{(0)}\frac{e_a}{m_a}\left\{(\vec{c_a}\times\vec{B})\mathbb{A}_a^{(1)}-[B^2\vec{c}_a-\vec{B}(\vec{c}_a\cdot\vec{B})]\mathbb{A}_a^{(2)}\right\}$$

and

$$-\vec{c_a}\cdot\left\{\sum_j e_j\int d\vec{c_j}f_j^{(0)}\vec{\mathbb{A}}_j(\vec{c_j}\times\vec{B})\right\}=(\vec{c_a}\times\vec{B})\mathbb{G}_B^{(1)}-[B^2\vec{c_a}-\vec{B}(\vec{c_a}\cdot\vec{B})]\mathbb{G}_B^{(2)}$$

where

$$\mathbb{G}_B^{(1)}=\frac{1}{2}\sum_j e_j\int d\vec{c_j}f_j^{(0)}\mathbb{A}_j^{(1)}[c_j^2-\frac{1}{B^2}(\vec{c_j}\cdot\vec{B})^2] \tag{3.17a}$$

and

$$\mathbb{G}_B^{(2)}=\frac{1}{2}\sum_j e_j\int d\vec{c_j}f_j^{(0)}\mathbb{A}_j^{(2)}[c_j^2-\frac{1}{B^2}(\vec{c_j}\cdot\vec{B})^2] \tag{3.17b}$$

Therefore, Eq. (3.16a) yields,

$$\begin{aligned}f_a^{(0)}\left(\frac{m_ac_a^2}{2kT}-\frac{5}{2}\right)\vec{c_a}&=-f_a^{(0)}\frac{e_a}{m_a}\left\{(\vec{c_a}\times\vec{B})\mathbb{A}_a^{(1)}-[B^2\vec{c_a}-\vec{B}(\vec{c_a}\cdot\vec{B})\mathbb{A}_a^{(2)}]\right\}\\&+\frac{m_a}{\rho kT}f_a^{(0)}\left[(\vec{c_a}\times\vec{B})\mathbb{G}_B^{(1)}-[B^2\vec{c_a}-\vec{B}(\vec{c_a}\cdot\vec{B})]\mathbb{G}_B^{(2)}\right]\\&+f_a^{(0)}\left\{C(\vec{c_a}\mathbb{A}_a^{(1)})+C(\vec{c_a}\mathbb{A}_a^{(1)}+\vec{c_b}\mathbb{A}_b^{(1)})\right\}\\&+f_a^{(0)}\left\{C\left[(\vec{c_a}\times\vec{B})\mathbb{A}_a^{(2)}\right]+C\left[(\vec{c_a}\times\vec{B})\mathbb{A}_a^{(2)}+(\vec{c_b}\times\vec{B})\mathbb{A}_b^{(2)}\right]\right\}\\&+f_a^{(0)}\left\{\left\{C(\vec{c_a}\mathbb{A}_a^{(3)})+C(\vec{c_a}\mathbb{A}_a^{(3)}+\vec{c_b}\mathbb{A}_b^{(3)})\right\}\right\}\cdot\vec{B}\vec{B}\end{aligned} \tag{3.18}$$

and an identical equation for species b. The corresponding equations for $\mathbb{D}_a^{(i)}$ and $\mathbb{D}_b^{(i)}$ where $i=1,2,3$ have exactly the same structure except that the inhomogeneous term is the one appearing in Eq. (3.16b). Moreover, it is now clear from the three independent vectors $\vec{c_a}$, $\vec{c_a}\times\vec{B}$ and $(\vec{c_a}\cdot\vec{B})\vec{B}$ that we have three independent equations namely,

$$\begin{aligned}f_a^{(0)}\left(\frac{m_ac_a^2}{2kT}-\frac{5}{2}\right)\vec{c_a}&=f_a^{(0)}\frac{e_aB^2}{m_a}\mathbb{A}_a^{(2)}\vec{c_a}-f_a^{(0)}\frac{m_aB^2}{\rho kT}\mathbb{G}_b^{(2)}\vec{c_a}\\&+f_a^{(0)}\left\{C(\vec{c_a}\mathbb{A}_a^{(1)})+C(\vec{c_a}\mathbb{A}_a^{(1)}+\vec{c_b}\mathbb{A}_b^{(1)})\right\}\end{aligned} \tag{3.19a}$$

$$0 = -f_a^{(0)} \frac{e_a}{m_a} (\vec{c_a} \times \vec{B}) \mathbb{A}_a^{(1)} + f_a^{(0)} \frac{m_a}{\rho kT} (\vec{c_a} \times \vec{B}) \mathbb{G}_b^{(1)}$$

$$+f_a^{(0)} \left\{ C \left[\left(\vec{c_a} \times \vec{B} \right) \mathbb{A}_a^{(2)} \right] + C \left[(\vec{c_a} \times \vec{B}) \mathbb{A}_a^{(2)} + (\vec{c_b} \times \vec{B}) \mathbb{A}_b^{(2)} \right] \right\} \tag{3.19b}$$

and taking out the factor $\vec{B}\vec{B}$,

$$0 = -f_a^{(0)} \frac{e_a}{m_a} \vec{c_a} \mathbb{A}_a^{(2)} + f_a^{(0)} \frac{m_a}{\rho kT} \vec{c_a} \mathbb{G}_b^{(2)} + f_a^{(0)} \left\{ C(\vec{c_a} \mathbb{A}_a^{(3)}) + C(\vec{c_a} \mathbb{A}_a^{(3)} + \vec{c_b} \mathbb{A}_b^{(3)}) \right\} \tag{3.19c}$$

This is a set of three linear coupled integral equations for the three functions $\mathbb{A}_a^{(i)}$ characterizing $\varphi_a^{(1)}$. An analogous set exist for $\varphi_b^{(1)}$ and the same reasoning leads to the equations for $\mathbb{D}_a^{(i)}$ and $\mathbb{D}_b^{(i)}$ coefficients. Eq. (3.19b) may be also simplified by factoring out $\times \vec{B}$ so that it reads,

$$0 = -f_a^{(0)} \frac{e_a}{m_a} \vec{c_a} \mathbb{A}_a^{(1)} + f_a^{(0)} \frac{m_a}{\rho kT} \vec{c_a} \mathbb{G}_b^{(1)} + f_a^{(0)} \left\{ C(\vec{c_a} \mathbb{A}_a^{(2)}) + C(\vec{c_a} \mathbb{A}_a^{(2)} + \vec{c_b} \mathbb{A}_b^{(2)}) \right\} \tag{3.19b$'$}$$

We may reduce this system to only two equations by the following trick. We multiply Eq. (3.19c) by B^2 and add it to Eq. (3.19a). The first two terms cancell out yielding

$$f_a^{(0)} \left(\frac{m_a c_a^2}{2kT} - \frac{5}{2} \right) \vec{c_a} = f_a^{(0)} \left\{ C(\vec{c_a} R_a) + C(\vec{c_a} R_a + \vec{c_b} R_b) \right\} \tag{3.20a}$$

where $R_a = \mathbb{A}_a^{(1)} + B^2 \mathbb{A}_a^{(3)}$. But this equation is identical to that obtained for an ordinary mixture so that the form the function R_i $(i = a, b)$ is already known. We need not worry any more about its solution. On the other hand if we multiply Eq. (3.19b) by iB and add the result to Eq. (3.19a) we get that[3]

$$f_a^{(0)} \left(\frac{m_a c_a^2}{2kT} - \frac{5}{2} \right) \vec{c_a} = f_a^{(0)} \frac{m_a}{\rho kT} (iB) \vec{c_a} G$$

$$-f_a^{(0)} \frac{e_a}{m_a} \vec{c_a} (iB) \mathbb{A}_a + f_a^{(0)} \left\{ C(\vec{c_a} \mathbb{A}_a) + C(\vec{c_a} \mathbb{A}_a + \vec{c_b} \mathbb{A}_b) \right\} \tag{3.20b}$$

where

$$\begin{aligned} \mathbb{A}_i &= \mathbb{A}_i^{(1)} + iB\mathbb{A}_i^{(2)} \\ G &= \mathbb{G}_B^{(1)} + iB\mathbb{G}_B^{(2)} \end{aligned} \tag{3.21}$$

---

[3] $B^2 = iB(-iB)$.

Eq. (3.20b) has a completely different structure as that of standard linear integral equations encountered in kinetic theory due to the appearance of the iB terms in the right hand side. Its solution remains as the main challenge of this section.

Notice should be made of the fact that an identical procedure leads to the equations for the $\mathbb{D}$ coefficients. The final result is that

$$\frac{n_a}{n} f_a^{(0)} \vec{c}_a = f_a^{(0)} \Big\{ C(\vec{c}_a(\mathbb{D}_a^{(1)} + B^2 \mathbb{D}_a^{(3)})) + C(\vec{c}_a(\mathbb{D}_a^{(1)} + B^2 \mathbb{D}_a^{(3)})) + \vec{c}_b(\mathbb{D}_b^{(1)} + B^2 \mathbb{D}_b^{(3)})) \Big\} \tag{3.22a}$$

and

$$\frac{n_a}{n} f_a^{(0)} \vec{c}_a = f_a^{(0)} \frac{m_a}{\rho k T} (iB) \vec{c}_a K - f_a^{(0)} \frac{e_a}{m_a} (iB) \vec{c}_a \mathbb{D}_i + f_a^{(0)} \left\{ C(\vec{c}_a \mathbb{D}_a) + C(\vec{c}_a \mathbb{D}_a + \vec{c}_b \mathbb{D}_b) \right\} \tag{3.22b}$$

where

$$\begin{aligned} \mathbb{D}_i &= \mathbb{D}_i^{(1)} + iB \mathbb{D}_i^{(2)} \\ K &= K_B^{(1)} + iB K_B^{(2)} \end{aligned} \tag{3.23}$$

and

$$K_B^{(1)} = \frac{1}{2} \sum_j e_j \int d\vec{c}_j f_j^{(0)} \mathbb{D}_j^{(1)} \left[c_j^2 - \frac{1}{B^2} (\vec{c}_j \cdot \vec{B})^2 \right] \tag{3.24a}$$

$$K_B^{(2)} = \frac{1}{2} \sum_j e_j \int d\vec{c}_j f_j^{(0)} \mathbb{D}_j^{(2)} \left[c_j^2 - \frac{1}{B^2} (\vec{c}_j \cdot \vec{B})^2 \right] \tag{3.24b}$$

The question is now how can we solve Eqs. (3.20b) and (3.22b) which as mentioned, are of a new type in kinetic theory. Nevertheless, before we do so we must see if $\varphi_i^{(1)}$ has no further restrictions imposed on it by the subsidiary conditions, Eqs. (3.11). Omitting the grad$\vec{u}$ term, substitution of Eq. (3.13) into Eq. (3.11) yields,

$$\sum_i m_i \int f_i^{(0)} \vec{\mathbb{A}}_i \left\{ \begin{array}{c} 1 \\ \vec{c}_i \\ \frac{1}{2} c_i^2 \end{array} \right\} d\vec{c}_i \cdot \text{ grad } \ln T +$$

$$\sum_i m_i \int f_i^{(0)} \vec{\mathbb{D}}_i \left\{ \begin{array}{c} 1 \\ \vec{c}_i \\ \frac{1}{2} c_i^2 \end{array} \right\} d\vec{c}_i \cdot \vec{d}_{ij} = 0$$

Since the two terms are similar, we fix our attention in the first one. Using the expression for $\vec{\mathbb{A}}_i$ we have for the non-vanishing integrals that,

$$\sum_i m_i \left\{ \int f_i^{(0)} \mathbb{A}_i^{(1)} \vec{c}_i \vec{c}_i d\vec{c}_i \cdot \text{ grad } \ln T + \int f_i^{(0)} \mathbb{A}_i^{(2)} \vec{c}_i \vec{c}_i d\vec{c}_i \cdot (\vec{B} \times \text{grad } \ln T) + \right.$$

$$\left. \int f_i^{(0)} \mathbb{A}_i^{(3)} \vec{c}_i \vec{c}_i d\vec{c}_i \vec{B} (\vec{B} \cdot \text{grad } \ln T) \right\} = 0$$

Since $\vec{c}_i \vec{c}_i = \overleftrightarrow{c_a{}^0 c_a} + \frac{1}{3} c_i^2 \mathbb{I}$ and noticing that all integrals with the symmetric traceless part vanish, we are left with three conditions,

$$\sum_i m_i \int f_i^{(0)} \mathbb{A}_i^{(1)} c_i^2 d\vec{c}_i \cdot \text{grad } \ln T = 0$$

$$\sum_i m_i \int f_i^{(0)} \mathbb{A}_i^{(2)} c_i^2 d\vec{c}_i \mathbb{I} \cdot (\vec{B} \times \text{grad } \ln T) = 0$$

$$\sum_i m_i \int f_i^{(0)} \mathbb{A}_i^{(3)} c_i^2 d\vec{c}_i B^2 \text{grad } \ln T = 0$$

provided $\vec{B}$ is taken, say along the z-axis, $\vec{B} = \hat{k} B$. Only then $\vec{B}(\vec{B} \cdot \text{grad } T) = B^2 \text{grad } T$. This assumption is unnecessary but shall be kept for didactical reasons. In fact, let $\hat{h}$ be a unit vector along $\vec{B} = (B_x, B_y, B_z)$. Multiply the first condition by $\hat{h}$ so that $\hat{h} \cdot \text{grad } T$ is the parallel component of grad T along $\hat{h}$. The same operation with the third condition yields $\vec{B}(\vec{B} \cdot \hat{h}) = B^2(\hat{h} \cdot \text{grad } T) h$ so adding both we get

$$\sum_i m_i \int f_i^{(0)} (\mathbb{A}_i^{(1)} + B^2 \mathbb{A}_i^{(3)}) c_i^2 d\vec{c}_i \hat{h} (\hat{h} \cdot \text{grad } T) = 0$$

from which the first of Eqs. (3.25) follows at once. Now multiply the first Eq. vectorially by $\hat{h}$ and add to the second one multiplied by i. Since $\mathbb{I} \cdot \hat{h} = \hat{h}$ we get

$$\sum_i m_i \int f_i^{(0)} (\mathbb{A}_i^{(1)} + iB \mathbb{A}_i^{(2)}) c_i^2 d\vec{c}_i (\hat{h} \times \text{grad } T) = 0$$

from which the second of Eqs. (3.25) follows at once. In general $\hat{h} = \hat{\imath} \cos\theta + \hat{\jmath} \cos\phi \sin\theta + \hat{k} \sin\phi \sin\theta$ in a cartesian coordinate system, thus emphasizing that in general $\vec{B}$ is in any arbitrary direction. Nevertheless by choosing

z-axis along $\vec{B}$ many algebraic operations are simplified but this has to be kept in mind specially when discussing magnetohydrodynamics.
So for $R_i = \mathbb{A}_i^{(1)} + B^2 \mathbb{A}_i^{(3)}$ and for $\mathbb{A}_i = \mathbb{A}_i^{(1)} + iB\mathbb{A}_i^{(2)}$ we finally get that the subsidiary conditions are:

$$\begin{aligned} \textstyle\sum_i m_i \int f_i^{(0)} R_i c_i^2 d\vec{c}_i = 0 \\ \textstyle\sum_i m_i \int f_i^{(0)} \mathbb{A}_i c_i^2 d\vec{c}_i = 0 \end{aligned} \tag{3.25}$$

and similar relations for the $\mathbb{D}_i^{(j)}$ functions. Eqs (3.25) follow also if as mentioned, $\vec{B} = \hat{k}B$ so that $\vec{B}(\vec{B} \cdot \text{grad } T) = B^2 \frac{\partial T}{\partial z} \hat{k}$.

The last step in this methodology is to propose convenient series expansions for the unknown functions in terms of a complete set of ortho-normal functions. For this purpose we select, as is now familiar in kinetic theory, the Sonine (Laguerre) polynomials $S_p^{(m)}$. Recall briefly that [2],

$$S_p^{(m)} = \sum_{r=0}^{p} \frac{(-x)^r (m+p)_{p-r}}{r!(p-r)!}$$

where

$$(m+p)_q = \prod_{s=0}^{q} (m+p+s)$$

so we get that for the first terms,

$$S_m^{(0)}(x) = 1 \; ; \qquad S_m^{(1)}(x) = -x + m + 1 \; ;$$

$$S_m^{(2)}(x) = \frac{1}{2}(m+1)(m+2) - (m+2)x + \frac{x^2}{2!}$$

and

$$\int_0^\infty e^{-x} S_m^{(p)}(x) S_m^{(q)}(x) x^m dx = \frac{\Gamma(m+p+1)}{p!} \delta_{pq} \tag{3.26}$$

are a few properties of these polynomials. We now propose that for species a, b,

$$\mathbb{A}_j^{(i)} = \sum_{m=0}^{\infty} (a_j^{(i)})^{(m)} S_{\frac{3}{2}}^{(m)}(c_j^2) \qquad \begin{array}{l} j = a, b \\ i = 1, 2, 3 \end{array} \tag{3.27}$$

where the coefficients $(a_j^{(i)})^{(m)}$ are functions of the scalars B^2, n, T, etc. A similar expression holds also for $\mathbb{D}_j^{(i)}$ which we shall not write down.

In the case of the functions $\mathbb{A}_i$ and $\mathbb{D}_i$ introduced in Eqs. (3.21) and (3.23), we shall write

$$\mathbb{A}_j^{(i)} = \mathbb{A}_j^{(1)} + iB\mathbb{A}_j^{(2)} = \sum_{m=0}^{\infty} \alpha_j^{(m)} S_{\frac{3}{2}}^{(m)}(c_j^2) \tag{3.28}$$

with

$$\alpha_j^{(m)} = (a_j^{(1)})^{(m)} + iB(a_j^{(2)})^{(m)}$$

and similar equations for $\mathbb{D}_i$ and $d_j^{(m)}$ respectively.

Eqs. (3.27) and (3.28) are very useful to considerably reduce both, the structure of the integral equations (3.20b) and (3.22b), as well as the subsidiary conditions expressed by Eqs. (3.25). In fact introducing the dimensionless velocity

$$\vec{\omega_i} = \sqrt{\frac{m_i}{2kT}}\vec{c_i} \tag{3.29}$$

noticing that $f_i^{(0)} = n_i \left(\frac{m_i}{2\pi kT}\right)^{\frac{3}{2}} e^{-\omega_i^2}$ and using the orthogonality condition for the Sonine polynomials, Eq. (3.26) one readily finds that Eqs. (3.25) reduce simply to

$$\sum_{i=a}^{b} n_i \left(a_{(i)}^{(1)(0)} + B^2 a_i^{(3)(0)}\right) = 0 \tag{3.30a}$$

$$\sum_{i=a}^{b} n_i a_{(i)}^{(0)} = 0 \tag{3.30b}$$

thus implying that all coefficients $a_{(i)}^{(m)}$ for $m > 0$ are not restricted. By the same argument, it is readily seen that,

$$G_{(B)}^{(1)} = \sum_j \frac{e_j n_j}{m_j} kT a_{(j)}^{(1)(0)} \tag{3.31a}$$

and

$$G_{(B)}^{(2)} = \sum_j \frac{e_j n_j}{m_j} kT a_{(j)}^{(2)(0)} \tag{3.31b}$$

and using Eq. (3.28),

$$G = \sum_j \frac{e_j n_j}{m_j} kT a_{(j)}^{(0)} \tag{3.32}$$

Analogously,

$$K = \sum_j \frac{e_j n_j}{m_j} kT d_{(j)}^{(0)} \tag{3.33}$$

With these results we can now begin to tackle the two immediate remaining questions namely, the evaluation of the physical fluxes $\vec{J}_a$ and $\vec{J}_q$ involving diffusive phenomena, heat conduction and corresponding cross effects. According to Eq. (2.15) the calculation of the conduction current requires evaluation of $\vec{J}_a$ which also involves, through $\vec{d}_{ab}$, mechanical and thermal diffusion. Therefore chapter 4 will be devoted to the computation of the physical fluxes (or currents).

Bibliography

[1] P. Goldstein and L. García-Colín, *J. Non-equilib. Thermodyn.* **30**, 173 (2005).

[2] See references in Chap. 1. Specially Ref. [8] part III.

Chapter 4

Calculation of the Currents

4.1 Diffusion Effects

We begin with calculating explicitly the expression for the mass current namely,

$$\vec{J}_a = m_a \int \vec{c}_a f_a^{(0)} \varphi_a^{(1)} d\vec{c}_a$$

which according to Eq. (3.13), omitting grad $\vec{u}$ terms yields

$$\vec{J}_a = m_a \int \vec{c}_a f_a^{(0)} \left\{ \vec{\mathbb{A}}_a \cdot \text{grad } \ln T + \vec{\mathbb{D}}_a \cdot \vec{d}_{ab} \right\} d\vec{c}_a$$

where $\vec{\mathbb{A}}_a$ and $\vec{\mathbb{D}}_a$ as well, have the form expressed in Eq. (3.15'). The next step is obvious. We introduce Eq. (3.15') into $\vec{J}_a$, use (3.27) and notice that the three integrals appearing in the resulting expression are

$$\int f_a^{(0)} \sum_{m=0}^{\infty} a^{(m)(i)} S_{\frac{3}{2}}^{(m)} \vec{c}_a \vec{c}_a d\vec{c}_a = \frac{n_a}{m_a} k T a^{(0)(i)} \mathbb{I}$$

when use is made of the fact that the contribution arising from the symmetric traceless part $\overleftrightarrow{c_a {}^0 c_a}$ vanishes and we resort to Eq. (3.26). Here $\mathbb{I}$ is the unit tensor.

L.S. García-Colín, L. Dagdug, *The Kinetic Theory of a Dilute Ionized Plasma*, Springer Series on Atomic, Optical, and Plasma Physics 53

Using this result and Eq. (3.13) in the definition of $\vec{J}_a$ we get that

$$\vec{J}_a = n_a kT \left\{ a_a^{(1)(0)} \text{ grad } \ln T + a_a^{(2)(0)} \text{ grad } \ln T \times \vec{B} + \right.$$

$$\left. \left(\vec{B} \cdot \text{ grad } \ln T\right) \vec{B} a_a^{(3)(0)} + \vec{d}_{ab} d_a^{(1)(0)} + \left(\vec{d}_{ab} \times \vec{B}\right) d_a^{(2)(0)} + \left(\vec{B} \cdot \vec{d}_{ab}\right) \vec{B} d_a^{(3)(0)} \right\} \tag{4.1}$$

with an analogous expression for $\vec{J}_b$ $(= -\vec{J}_a)$ remembering that $\vec{d}_{ab} = -\vec{d}_{ba}$. Eq. (4.1) is a very eloquent result: to know $\vec{J}_a$ we do not need the full information contained in the infinite series (3.27) but only six coefficients, three for each of the thermodynamic forces. This will considerably simplify the solutions to the integral equations derived in the previous section. Notice also how the magnetic field strongly influences the diffusive, thermal and mechanical effects giving rise to the mass flux $\vec{J}_a$. This is readily appreciated examining Eq. (3.10) which even in the absence of an external field has two contributions, the first arising from the standard diffusive first two terms and a second one arising from the term proportional to $\vec{E}'$ which includes the effect of the magnetic field. In order to understand the full character of Eq. (4.1) let us examine the structure of its terms, one by one in a three dimensional space assuming that the direction of the magnetic field is taken along the z-axis. Thus,

$$\vec{B} = B\hat{\mathrm{k}}$$

$\hat{\mathrm{k}}$ being a unitary vector along z-axis. Now we rearrange Eq. (4.1) conveniently. The third term is simply $B^2 \frac{\partial T}{\partial z}$ which may be combined with the corresponding contribution in the first term to yield $\left(a_a^{(1)(0)} + a_a^{(3)(0)} B^2\right) \frac{\partial T}{\partial z}$. Also, $\vec{B} \times$ grad $T = B\left(\hat{\mathrm{j}}\frac{\partial T}{\partial x} - \hat{\mathrm{i}}\frac{\partial T}{\partial y}\right)$, $\hat{\mathrm{j}}$ and $\hat{\mathrm{i}}$ being unit vectors along the y and x axis, respectively. With a similar procedure for vector $\vec{d}_{ab}$, we are readily lead to the following expression,

$$\vec{J}_a = n_a k(\mathrm{a}_a^{(1)(0)} + \mathrm{a}_a^{(3)(0)} B^2)(\text{grad } T)_\parallel + n_a k \mathrm{a}_a^{(1)(0)} (\text{grad } T)_\perp +$$

$$n_a k \mathrm{a}_a^{(2)(0)} B (\text{grad } T)_s + n_a kT \left\{ \left(d_a^{(1)(0)} + B^2 d_a^{(3)(0)}\right) \left(\vec{d}_{ab}\right)_\parallel \right\} +$$

$$n_a kT d_a^{(1)(0)} (\vec{d}_{ab})_\perp + n_a kT d_a^{(2)(0)} B (\vec{d}_{ab} \times \hat{\mathrm{k}}) \tag{4.2}$$

where, for simplicity $(\text{grad } T)_\parallel \equiv \hat{\mathrm{k}}\frac{\partial T}{\partial z}$, $(\text{grad } T)_\perp \equiv \hat{\mathrm{i}}\frac{\partial T}{\partial x} + \hat{\mathrm{j}}\frac{\partial T}{\partial y}$ and $(\text{grad } T)_s \equiv -\hat{\mathrm{i}}\frac{\partial T}{\partial y} + \hat{\mathrm{j}}\frac{\partial T}{\partial x}$ (see Fig. (4.1)). Also $\left(\vec{d}_{ab} \cdot \hat{\mathrm{k}}\right) \hat{\mathrm{k}} \equiv \left(\vec{d}_{ab}\right)_\parallel \hat{\mathrm{k}}$ is the total diffusion

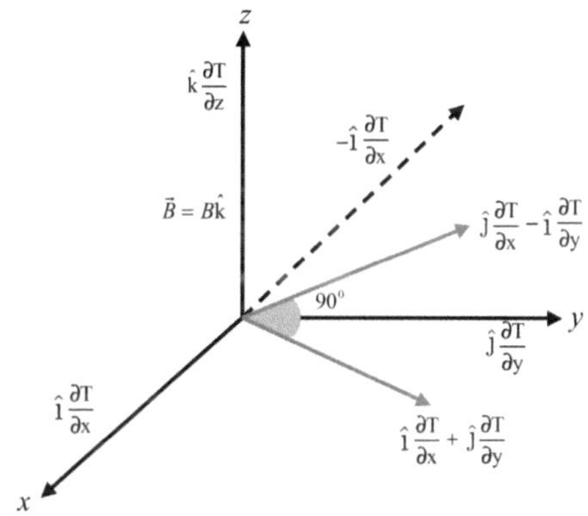

Figure 4.1: The different vectors which result from terms $B\times$grad T and $B\cdot$grad $T = B\frac{\partial T}{\partial z}$. The fluxes can be parallel to $\vec{B}$, perpendicular to B (x-y plane) and perpendicular to this plane and to the z-axis.

along the B direction and $\left(\vec{d}_{ab}\right)_{\perp} = \hat{\imath}\,(d_a)_x + \hat{\jmath}\,(d_a)_y$. The term $\hat{k}\times\vec{d}_{ab}$ merits a closer look to be discussed later. Thus we have in general matter (and charge) flowing in the direction parallel to the magnetic field, in the $(x,\,y)$ plane in a direction perpendicular to the field and in a third direction, the $(-x,\,y)$ plane which is both perpendicular to the field and the $(x,\,y)$ plane. When $\vec{B}=\vec{0}$ the first three terms simply yield an expression for the ordinary thermal diffusion term, the Soret effect.

$$\left(\vec{J}_a\right)_{Soret} = n_a k a_a^{(1)(0)} \text{grad } T$$

also in the absence of an electric field.

To exploit Eq. (4.2) let us study the flow of electrical charge according to Eq. (2.15). For this purpose we write

$$\vec{d}_{ab} = \vec{d}_{ab}^{(0)} - \frac{n_a n_b}{p\rho}(m_b e_a - m_a e_b)\vec{E}' \tag{4.3}$$

where $\vec{d}_{ab}^{(0)}$ is the standard non electromagnetic part of $\vec{d}_{ab}$ (cf. Eq. (3.10)). Since $\left(\vec{u}\times\vec{B}\right)_z = 0$,

$$\vec{E}' = \vec{E} + \left(\vec{u}\times\vec{B}\right) = (E_x + u_y B)\,\hat{\imath} + (E_y - u_x B)\,\hat{\jmath} + E_z\hat{k}$$

where

$$\hat{\mathrm{k}} \times \vec{E}' = (E_x + u_y B)\hat{\jmath} - (E_y - u_x B)\hat{\imath}$$

so calling

$$\frac{n_a^2 n_b kT}{p\rho}(m_b e_a - m_a e_b) \equiv \theta \tag{4.4}$$

and re-arranging terms, the electrical conduction current for species a is given by

$$\vec{J}_a^{(c)} = -\theta(d_a^{(1)(0)} + B^2 d_a^{(3)(0)})E_z\hat{\mathrm{k}} - d_a^{(1)(0)}\theta\left\{(E_x + u_y B)\hat{\imath} + (E_y - u_x B)\hat{\jmath}\right\} -$$

$$d_a^{(2)(0)}\theta B\left\{(E_x + u_y B)\hat{\jmath} - (E_y - u_x B)\hat{\imath}\right\} \tag{4.5}$$

Using Eqs. (2.15), (4.4) and (4.5) we see that we have three electrical conductivities namely,

$$\sigma_a^{\parallel} = \frac{n_a^2 n_b (m_a + m_b)^2}{m_a m_b p\rho} e^2 kT(d_a^{(1)(0)} + d_a^{(3)(0)} B^2) \tag{4.6a}$$

corresponding to the current flow parallel to the magnetic field if $E_z \neq 0$; a current in the (x,y) plane whose conductivity is,

$$\sigma_a^{\perp} = \frac{n_a^2 n_b (m_a + m_b)^2}{m_a m_b p\rho} e^2 kT d_a^{(1)(0)} \tag{4.6b}$$

and the one corresponding to the current perpendicular both to the (x, y) plane and the direction of B, namely

$$\sigma_a^{s} = \frac{n_a^2 n_b (m_a + m_b)^2}{m_a m_b p\rho} e^2 kT d_a^{(2)(0)} \tag{4.6c}$$

If $B = 0$, $\sigma_a^{\parallel} = \sigma_a^{\perp}$ and the current flows along the direction of $\vec{E}$. Further, if the electric field is fixed in a certain direction with respect to $\vec{B}$ the respective flows can be easily derived from the general expression, Eq. (4.5). We further insist that to explicitly compute the conductivities, only three quantities are needed, the coefficients $d_a^{(i)(0)}$ where $i = 1, 2, 3$ (since $\vec{J}_a = -\vec{J}_b$). The several cross effects which arise from Eq. (4.2) when $d_{ab}^{(0)}$ is taken into account will be left out for a later discussion (see Chapter 6).

4.2 Flow of Heat

To study this quantity we use the same definition as in the no magnetic field case. The standard definition for a mixture, according to classical irreversible thermodynamics is [3, 7],

$$\frac{1}{kT}\vec{J'}_q = \frac{1}{kT}\vec{J}_q - \frac{5}{2}\left(\frac{\vec{J}_a}{m_a} + \frac{\vec{J}_b}{m_b}\right) \tag{4.7}$$

so that using Eqs. (2.12) and (2.23) we get that

$$\frac{1}{kT}\vec{J'}_q = \sum_{i=a}^{b}\int\left(\frac{m_i c_i^2}{2kT} - \frac{5}{2}\right)\vec{c}_i f_i d\vec{c}_i$$

and substituting the explicit form for f_i, Eq. (3.13),

$$\frac{1}{kT}\vec{J'}_q = \sum_{i=a}^{b}\int\left(\frac{m_i c_i^2}{2kT} - \frac{5}{2}\right)\vec{c}_i f_i^{(0)}\left[\vec{\mathbb{A}}_a \cdot \text{ grad}\ln T \pm \vec{\mathbb{D}}_a \cdot \vec{d}_{ij}\right] d\vec{c}_i$$

where the $\pm$ sign appears since $\vec{d}_{ij} = -\vec{d}_{ji}$. If we now make use of. Eqs. (3.15a,b) for $\vec{\mathbb{D}}_a$ and $\vec{\mathbb{A}}_a$ we readily see that all six integrals that appear in the resulting expression are of the form

$$\sum_{i=a}^{b}\int\left(\frac{m_i c_i^2}{2kT} - \frac{5}{2}\right)\frac{1}{3}c_i^2 f_i^{(0)}\sum_{m=0}^{\infty} a_i^{(j)(m)} S_{\frac{3}{2}}^{(m)}(c_i^2)d\vec{c}_i = -\frac{5}{2}kT\sum_{i=a}^{b}\frac{n_i}{m_i}a_i^{(j)(1)},$$

after Eq. (3.26) has been used and integrations performed using Eq. (3.27). Thus we arrive at the result that

$$\vec{J'}_q = -\frac{5}{2}k^2T\left\{\sum_{i=a}^{b}\frac{n_i}{m_i}\left[a_i^{(1)(1)}\text{ grad } T + a_i^{(2)(1)}\text{ grad } T \times \vec{B} + a_i^{(3)(1)}B^2\left(\text{ grad } T\right)_{\parallel}\right]\right\}$$

$$-\frac{5}{2}(kT)^2\left\{\sum_{i=a}^{b}\frac{n_i}{m_i}(\pm)\left[d_i^{(1)(1)}\vec{d}_{ab} + d_i^{(2)(1)}\vec{B}\times\vec{d}_{ab} + d_i^{(3)(1)}B^2\left(\vec{d}_{ab}\right)_{\parallel}\right]\right\} \tag{4.8}$$

where $\pm$ means that a minus sign appears when $i = b$ to account for the property that $\vec{d}_{ab} = -\vec{d}_{ba}$. Eq. (4.8) contains all possible contributions to the flow of heat namely those coming from grad$T$ modified by the presence of the magnetic field together with all the cross effect provided by the purely diffusive term $d_{ab}^{(0)}$ and the electromagnetic contributions coming from $\theta \vec{E}'$, θ being defined in Eq. (4.4). Notice that if $\vec{B} = \vec{0}$ the ordinary thermal conductivity is simply given by

$$\kappa = \frac{5}{2} k^2 T \sum_i \frac{n_i}{m_i} a_i^{(1)(1)} \tag{4.9}$$

whose explicit values will be discussed later. Using the same notation as in the previous case with $\vec{B} = B\hat{\mathrm{k}}$, Eq. (4.8) may be written as follows,

$$\begin{aligned}\vec{J'}_q = -\frac{5}{2} k^2 T \sum_{i=a}^{b} \frac{n_i}{m_i} \Big\{ &(a_i^{(1)(1)} + B^2 a_i^{(3)(1)})(\text{ grad } T)_\parallel + a_i^{(1)(1)}(\text{ grad } T)_\perp \\ &+ a_i^{(2)(1)} B(\text{ grad} T)_s + (\pm) T[(d_i^{(1)(1)} + B^2 d_i^{(3)(1)})(\vec{d}_{ab})_\parallel + \\ &d_i^{(1)(1)}(\vec{d}_{ab})_\perp + d_i^{(2)(1)}(\vec{d}_{ab} \times \vec{B})] \Big\}\end{aligned} \tag{4.10}$$

Eq. (4.10) shows there are three different thermal conductivities. One corresponding to flow in the direction of the field

$$\kappa_\parallel = \frac{5}{2} k^2 T \sum_{i=a}^{b} \frac{n_i}{m_i} \left(a_i^{(1)(1)} + B^2 a_i^{(1)(3)} \right) \tag{4.11}$$

A second one corresponds to the heat flux in the xy plane, thus the thermal current flowing along the direction of the magnetic field and the one flowing along the (x, y) plane are not the same. Only when $\vec{B} = \vec{0}$, there is a unique current along the direction of grad T,

$$\kappa_\perp = \frac{5}{2} k^2 T \sum_{i=a}^{b} \frac{n_i}{m_i} a_i^{(1)(1)} \tag{4.12}$$

so that $\kappa_\parallel = \kappa_\perp$.

The third one is a thermal current flowing in the $(-x, y)$ plane, perpendicular both to the z axis and the (x, y) plane with a thermal conductivity given by

$$\kappa_s = \frac{5}{2} k^2 T \sum_{i=a}^{b} \frac{n_i}{m_i} a_i^{(2)(1)} B \tag{4.13}$$

which is the so-called Righi-Leduc effect (see Ref. [1] p. 337 and Refs. [2], [3]).

The Dufour effect arising from the term $d_{ab}^{(0)}$ as well as the electromagnetic contribution to the heat current can be easily extracted from the second line in Eq. (4.8) but we shall leave those details for a discussion in Chapter 6. What is altogether important to stress is that only the six coefficients $a_j^{(i)(1)}$ and $d_j^{(i)(1)}$ where $i = 1, 2, 3$ and $j = a, b$ are needed to explicitly compute any transport property associated with the heat transport.

With all this information in our hands it is now convenient to tackle the most difficult part of this work, namely the solution to the linear integral equations (3.20b) and (3.22b) which as said before, are not the standard type of equations appearing in the simple one component system in kinetic theory.

Bibliography

[1] S. Chapman and T. G. Cowling, loc. cit. Chap. 1.

[2] D. Miller, *Chem. Rev.* **60**, 15 (1960).

[3] R. Haase, *Thermodynamics of Irreversible Processes*; Addison-Wesley Publ. Co., Reading, Mass (1969).

[4] R. Balescu, *Transport Processes in Plasma Vol. I, Classical Transport*; North Holland Publ. Co., Amsterdam (1988).

[5] L. García-Colín, A. Sandoval-Villalbazo and A. L. García-Perciante, *Physics of Plasmas* 14, 012305 (2007); ibid **14**, 089901 (2007).

[6] L. S. García-Colín, A. L. García-Preciante, A. Sandoval-Villalbazo, *J. Non-equilib. Thermodyn.* **32**, 379 (2007).

[7] S. R. de Groot and P. Mazur, *Non-equilibrium Thermodynamics*, Dover Publications, Mineola N. Y. (1984).

Chapter 5

Solution of the Integral Equations

As pointed out before, Eqs. (3.20a) and (3.22a) are of the standard type of linear integral equations encountered in the kinetic theory of neutral systems so we need not worry about their solutions at all. We shall give a summary of their main properties in Appendix C. Here we wish to deal with Eqs. (3.20b) and (3.22b) whose structure becomes quite complicated due to the presence of the three magnetic field dependent terms appearing in their right hand side. Since they essentially differ only in structure by their inhomogeneous term, let us fix our attention to one of them namely, Eq. (3.20b) whose solution will yield the values of the "a" coefficients required to compute the three thermal conductivities. We shall seek a solution to this equation using a variational method which is apparently due to Davison [1]. We take $\mathfrak{I}_i$ $(i = a, b)$ as a trial function for $\mathbb{A}$ and construct a functional $\mathfrak{D}(\mathfrak{I}_i)$ by multiplying the full equation (3.20b) by $\mathfrak{I}_i\vec{c}_i$ integrating over $\vec{c}_i$ and summing over i. Thus,

$$\mathfrak{D}(\mathfrak{I}_i) = \sum_{i=a}^{b} \int d\vec{c}_i \mathfrak{I}_i \vec{c}_i \cdot \left[-f_i^{(0)} \left(\frac{m_i c_i^2}{2kT} - \frac{5}{2} \right) \vec{c}_i - f_i^{(0)} \frac{e_i}{m_i} (iB) \vec{c}_i \right.$$
$$\left. + f_i^{(0)} \frac{m_i}{\rho kT} (iB) G \vec{c}_i \right] + \sum_{i=a}^{b} \int d\vec{c}_i \vec{c}_i \cdot \left[f_i^{(0)} \left(C(\vec{c}_i \mathfrak{I}_i) \right) + C \left(\vec{c}_i \mathfrak{I}_i + \vec{c}_j \mathfrak{I}_j \right) \right] \quad (5.1a)$$

Now, from the definition of $\mathbb{A} = \mathbb{A}^{(1)} + iB\mathbb{A}^{(2)}$ and Eqs. (3.17a-b) the third term reads as

$$\sum_{i=a}^{b} \int d\vec{c}_i \mathfrak{I}_i \vec{c}_i \vec{c}_i f_i^{(0)} \frac{m_i}{kT} (iB) \left\{ \frac{1}{2} \sum_{j} \int dc_i f_j^{(0)} \mathfrak{I}_i \left[c_j^2 - \frac{1}{B^2} (\vec{c}_i \cdot \vec{B})^2 \right] \right\}$$

L.S. García-Colín, L. Dagdug, *The Kinetic Theory of a Dilute Ionized Plasma*, Springer Series on Atomic, Optical, and Plasma Physics 53

but the curly bracket no longer depends on $\vec{c}_i$ so this expression is equal to

$$\frac{iB}{kT}\left\{\quad\right\}\frac{1}{3}\sum_j m_i\int dc_i\mathfrak{I}_i c_i^2 f_i^{(0)}=0$$

since $\mathfrak{I}_i$ must obey the subsidiary conditions, Eq. (3.25). Thus,

$$\mathfrak{D}(\mathfrak{I}_i)=\sum_{i=a}^{b}\int d\vec{c}_i\mathfrak{I}_i\vec{c}_i\cdot\left[-f_i^{(0)}\left(\frac{m_ic_i^2}{2kT}-\frac{5}{2}\right)\vec{c}_i-\frac{iBe_i}{m_i}f_i^{(0)}\mathfrak{I}_i\vec{c}_i\right]+$$

$$\left[\sum_{i=a}^{b}\int d\vec{c}_i f_i^{(0)}\left(C(\vec{c}_i\mathfrak{I}_i)\right)+C\left(\vec{c}_i\mathfrak{I}_i+\vec{c}_j\mathfrak{I}_j\right)\right]\vec{c}_i \tag{5.1b}$$

Now, we want to seek a solution which satisfies the extremal condition

$$\delta\mathfrak{D}(\mathfrak{I}_i)=0$$

consistent with Eqs. (3.25). To perform the variation we use the expression for the linearized kernel $C(\vec{c}_i\mathfrak{I}_i)$ and $C\left(\vec{c}_i\mathfrak{I}_i+\vec{c}_j\mathfrak{I}_j\right)$ so the last two terms in the above expression may be written as

$$\sum_{i,j}\int\cdots\int d\vec{c}_id\vec{c'}_jd\vec{c'}_if_i^{(0)}f_j^{(0)}g_{ij}\sigma(\vec{c}_i\vec{c}_j\to\vec{c'}_i\vec{c'}_j)\mathfrak{I}_i\cdot\vec{c}_i\cdot$$

$$\left[(\vec{c'}_i\mathfrak{I'}_i)+(\vec{c'}_j\mathfrak{I'}_j)-(\vec{c}_i\mathfrak{I}_i)-(\vec{c}_j\mathfrak{I}_j)\right]$$

the convention of notation being the same as in Eq. (2.4). In fact this expression arises simply from Eq. (2.4) substituting f_i by $f_i^{(0)}(1+\varphi_i^{(1)})$ and keeping terms linear in $\varphi_i^{(1)}$. Now, we perform the two transformations leading to the H theorem, exchange first i with j and next exchange $\vec{c}_i$ with $\vec{c'}_i$ and $\vec{c}_j$ with $\vec{c'}_j$ using the fact that σ satisfies Eq. (2.5). This yields for Eq. (5.1b),

$$\mathfrak{D}(\mathfrak{I}_i)=-\sum_i\int d\vec{c}_ic_i^2\left(\frac{m_ic_i^2}{2kT}-\frac{5}{2}\right)f_i^{(0)}\mathfrak{I}_i-iB\sum_i\frac{e_i}{m_i}\int d\vec{c}_if_i^{(0)}\mathfrak{I}_i^2c_i^2+$$

$$\frac{1}{4}\sum_{i,j}\int\cdots\int d\vec{c}_id\vec{c'}_jd\vec{c'}_if_i^{(0)}f_j^{(0)}g_{ij}\sigma\left[(\vec{c'}_i\mathfrak{I'}_i)+(\vec{c'}_j\mathfrak{I'}_j)-(\vec{c}_i\mathfrak{I}_i)-(\vec{c}_j\mathfrak{I}_j)\right]^2$$

where $[\quad]^2$ implies $[\quad]\cdot[\quad]$.

Now we carry out the variation of $\mathfrak{I}_i$ and get that

$$\delta\mathfrak{D}(\mathfrak{I}_i) = -\sum_i \int d\vec{c}_i c_i^2 \left(\frac{m_i c_i^2}{2kT} - \frac{5}{2}\right) f_i^{(0)} \delta\mathfrak{I}_i - 2iB \sum_i \frac{e_i}{m_i} \int d\vec{c}_i f_i^{(0)} \mathfrak{I}_i^2 c_i^2 \delta\mathfrak{I}_i +$$

$$\frac{1}{2} \sum_{i,j} \int \cdots \int d\vec{c}_i d\vec{c'}_j d\vec{c'}_i f_i^{(0)} f_j^{(0)} g_{ij} \sigma$$

$$\left[\quad\right] \left[(\vec{c'}_i \delta\mathfrak{I}'_i) + (\vec{c'}_j \delta\mathfrak{I}'_j) - (\vec{c}_i \delta\mathfrak{I}_i) - (\vec{c}_j \delta\mathfrak{I}_j)\right] = 0$$

Next we perform all the H-theorem operations again over this last terms so that the last bracket appears only as $\vec{c}_i \delta\mathfrak{I}_i$ so that

$$\delta\mathfrak{D}(\mathfrak{I}_i) = -\sum_i \int d\vec{c}_i c_i \delta\mathfrak{I}_i \left(\frac{m_i c_i^2}{2kT} - \frac{5}{2}\right) f_i^{(0)} \vec{c}_i - 2iB \frac{e_i}{m_i} f_i^{(0)} c_i^2 \mathfrak{I}_i +$$

$$\sum_j \int d\vec{c'}_i d\vec{c'}_j f_i^{(0)} f_j^{(0)} g_{ij} \sigma \vec{c}_i \cdot \left[\quad\right]) = 0$$

for all of $\mathfrak{I}_i$. Hence the resulting equation for $\mathfrak{I}_i$ is

$$f_i^{(0)} \left(\frac{m_i c_i^2}{2kT} - \frac{5}{2}\right) = -2iB\vec{c}_i \frac{e_i}{m_i} f_i^{(0)} c_i \mathfrak{I}_i + 2 f_i^{(0)} \left[C\left(\vec{c}_i \mathfrak{I}_i\right) + \right.$$

$$\left. C\left(\vec{c}_i \mathfrak{I}_i + \vec{c}_j \mathfrak{I}_j\right)\right] + 2\alpha f_i^{(0)} \vec{c}_i \tag{5.2}$$

where α is an undetermined parameter that does not depend on i and requires that $\mathfrak{I}_i$ satisfies Eq. (3.25). But we have shown in p. 51 that indeed this holds true since in the original Eq. (3.20b) the term with the G function vanishes so that we can assert that Eq. (5.2) is equivalent to Eq. (3.20b). This implies that $\mathfrak{I}_i$, a solution to Eq. (5.2) which yields a stationary condition for $\mathfrak{D}(\mathfrak{I}_i)$ is the appropriate solution to Eq. (3.20b). Notice further that if we take

$$\mathfrak{I}_i = \mathbb{A}_i + \mathcal{O}(\delta^2)$$

where δ is small,

$$\mathfrak{D}_{\mathfrak{i}}(\mathfrak{I}) = \mathfrak{D}(\mathbb{A}_i) + \mathcal{O}(\delta^2)$$

so that a not so good trial function gives already a reasonably good value for $\mathfrak{D}(\mathfrak{I}_i)$ which is what we are seeking for. Further, this now entitles us to fully use Eq. (3.27) and propose as a solution to Eq. (5.2), the Sonine expansion

$$\mathfrak{I}_i = \mathbb{A}_i = \sum_{m=0}^{M} a_i^{(m)} S_{\frac{3}{2}}^{(m)}(c_i) \tag{5.3}$$

so that to apply the variational procedure we must evaluate each term in Eq. (5.1a); the limit M in the sum, taken to label the order of approximations. Now

$$\sum_i \int f_i^{(0)} \left(\frac{m_i c_i^2}{2kT} - \frac{5}{2}\right) c_i^2 \sum_{m=0}^{M} a_i^{(m)} S_{\frac{3}{2}}^{(m)} d\vec{c}_i = \frac{15}{2} kT \sum_i \frac{n_i}{m_i} a_i^{(1)}$$

when Eq. (3.26) is used. The second term yields:

$$\frac{iBe_i}{m_i} \sum_i \int d\vec{c}_i c_i^2 \Im_i^2 f_i^{(0)} =$$

$$iB \sum_i \frac{e_i}{m_i} \int_0^\infty d\vec{c}_i c_i^2 \sum_{m=0}^{M} a_i^{(m)} S_{\frac{3}{2}}^{(m)}(c_i) \sum_{m'=0}^{M} a_i^{(m')} S_{\frac{3}{2}}^{(m')}(c_i) f_i^{(0)}$$

which after a transformation to the dimensionless velocity ω_i, setting $m = m'$ by the orthonormality property of Sonine polynomials, using Eq. (3.26), and carrying out the integration over ω_i yields,

$$\sum_i \int d\vec{c}_i c_i^2 \Im_i^2 f_i^{(0)} \frac{iBe_i}{m_i} = iBkT \sum_i \frac{e_i n_i}{m_i^2} \left(6(a_i^{(0)})^2 + 15(a_i^{(1)})^2\right)$$

The third term in (5.1a) vanishes due to the subsidiary conditions so that we are left with the expression for the linearized collision kernel. However, its adequate handling requires of some results concerning the properties of collision integrals discussed at length in the Appendix B. In fact with the same approximation as used for the second term, namely $M = 0, 1$ the full expression for the linearized collision kernels, reads

$$I \equiv \sum_{i,j} \int \cdots \int f_i^{(0)} f_j^{(0)} d\vec{v'}_i d\vec{v'}_j \sigma(\Omega) d\Omega g_{ij} \left\{ a_i^{(0)} \vec{c}_i - a_i^{(1)} \vec{c}_i (\omega_i^2 - \frac{5}{2}) \right\}$$
$$\left\{ a_i^{(0)} \vec{c'}_i - a_i^{(1)} \vec{c'}_i (\omega_i'^2 - \frac{5}{2}) + a_j^{(0)} \vec{c'}_j - a_j^{(1)} \vec{c'}_j (\omega_j'^2 - \frac{5}{2}) \right.$$
$$\left. - a_i^{(0)} \vec{c}_i + a_i^{(1)} \vec{c}_i (\omega_i^2 - \frac{5}{2}) - a_j^{(0)} \vec{c}_j + a_j^{(1)} \vec{c}_j (\omega_j^2 - \frac{5}{2}) \right\}$$

Exchanging subscripts i and j and subsequently setting $\vec{c}_i \to \vec{c'}_i$, $\vec{c}_j \to \vec{c'}_j$, the same transformation used in proving the H-theorem, and using Eqs. (B.1) and (B.2) of appendix B it follows at once that

$$I = -\frac{1}{2} \sum_{i,j} n_i n_j \left[G_{ij}, H_{ij}\right] \tag{5.4a}$$

where

$$G_{ij} = a_i^{(0)}\vec{c}_i - a_i^{(1)}\vec{c}_i(\omega_i^2 - \tfrac{5}{2}) + a_j^{(0)}\vec{c}_j - a_j^{(1)}\vec{c}_j(\omega_j^2 - \tfrac{5}{2})$$
$$H_{ij} = G_{ij} \tag{5.4b}$$

If we now define

$$K_i = a_i^{(0)}\vec{c}_i - a_i^{(1)}\vec{c}_i(\omega_i^2 - \tfrac{5}{2})$$
$$L_j = a_j^{(0)}\vec{c}_j - a_i^{(1)}\vec{c}_j(\omega_j^2 - \tfrac{5}{2}), \tag{5.5}$$

taking $M_i = K_i$, $M_j = L_j$ and using Eq. (B.4) together with Eq. (5.5) after some tedious algebra we find that

$$\begin{aligned} I = {} & 2n_a^2(a_a^{(1)})^2\left[\vec{c}_a(\omega_a^2 - \frac{5}{2}), \vec{c}_a(\omega_a^2 - \frac{5}{2})\right]_{aa} \\ & +2n_b^2(a_b^{(1)})^2\left[\vec{c}_b(\omega_b^2 - \frac{5}{2}), \vec{c}_b(\omega_b^2 - \frac{5}{2})\right]_{bb} \\ & +n_a n_b\Big\{(a_a^{(0)})^2\left[\vec{c}_a, \vec{c}_a\right]_{ab} - 2a_a^{(0)}a_a^{(1)}\left[\vec{c}_a, \vec{c}_a(\omega_a^2 - \frac{5}{2})\right]_{ab} \\ & +(a_a^{(1)})^2\left[[\vec{c}_a(\omega_a^2 - \frac{5}{2}), \vec{c}_a(\omega_a^2 - \frac{5}{2})\right]_{ab} \\ & +(a_b^{(0)})^2\left[\vec{c}_b, \vec{c}_b\right]_{ab} - 2a_a^{(0)}a_b^{(1)}\left[\vec{c}_b, \vec{c}_b(\omega_b^2 - \frac{5}{2})\right]_{ab} \\ & +(a_b^{(1)})^2\left[\vec{c}_b(\omega_b^2 - \frac{5}{2}), \vec{c}_b(\omega_a^2 - \frac{5}{2})\right]_{ab} + \\ & +2a_a^{(0)}a_b^{(0)}\left[\vec{c}_a, \vec{c}_b\right]_{ab} - 2a_a^{(0)}a_b^{(1)}\left[\vec{c}_b, \vec{c}_b(\omega_b^2 - \frac{5}{2})\right]_{ab} - 2a_b^{(0)}a_a^{(1)}\left[[\vec{c}_b, \vec{c}_a(\omega_a^2 - \frac{5}{2})\right]_{ab} + \\ & +2a_a^{(1)}a_b^{(1)}\left[\vec{c}_a(\omega_a^2 - \frac{5}{2}), \vec{c}_b(\omega_b^2 - \frac{5}{2})\right]_{ab}\Big\} \end{aligned} \tag{5.6}$$

where the twelve collision integrals appearing here are evaluated in Appendix D. In that appendix we see that all of them, when properly written and evaluated using the dimensionless velocity ω_i are related to a single collision integral $[\omega_a, \omega_a]_{ab} \equiv \varphi$, in the limit when $m_a >> m_b$. Writing I in terms of φ, introducing the appropriate factor $(\frac{2kT}{m_i})^{\frac{1}{2}}$ for each $\vec{c}_i$ in the collision term, and collecting all three terms composing Eq. (5.1a) we finally find that

$$\mathfrak{D}(\mathfrak{I}_i) = -\frac{15}{2}kT\left(\frac{n_a}{m_a}a_a^{(1)} + \frac{n_b}{m_b}a_b^{(1)}\right) + 3iBkT\sum_i \frac{e_i n_i}{m_i^2}\left(2(a_i^{(0)})^2 + 5(a_i^{(1)})^2\right)$$

$$+n_a n_b \varphi kT\Big\{\frac{1}{m_a}(a_a^{(0)})^2 - \frac{3}{m_a}a_a^{(0)}a_a^{(1)} + \frac{13}{4m_a}(a_a^{(1)})^2 + \frac{M_1}{m_b}(a_b^{(0)})^2$$

$$-\frac{3}{\sqrt{m_a m_b}}a_b^{(0)}a_b^{(1)} + \frac{15M_1}{2}(a_b^{(1)})^2 - 2\sqrt{\frac{M_1}{m_a m_b}}a_a^{(0)}a_b^{(0)} - \frac{3M_1^2}{m_b}a_a^{(0)}a_b^{(1)}$$

$$+3\sqrt{\frac{M_1}{m_a m_b}}a_b^{(0)}a_a^{(1)} - \frac{27M_1^{\frac{3}{2}}}{2\sqrt{m_a m_b}}a_a^{(1)}a_b^{(1)}\Big\}$$

$$+2kT\varphi\Big\{\frac{\sqrt{2}}{m_a}(a_a^{(1)})^2 n_a^2 + \frac{\sqrt{2M_1}}{m_b}(a_b^{(1)})^2 n_b^2\Big\} \tag{5.7}$$

which by the computational condition $\delta\mathfrak{D}(\mathfrak{I}) = 0$ will lead to a set of algebraic equations for the unknown coefficients $a_a^{(0)}$, $a_b^{(0)}$, $a_a^{(1)}$ and $a_b^{(1)}$. Here, $M_1 = \frac{m_a}{m_a+m_b}$, φ is known and its value is given in Appendix C. We still have to introduce the subsidiary condition given in Eq. (3.30b) namely,

$$n_a a_a^{(0)} + n_b a_b^{(0)} = 0$$

Nevertheless before taking the variation of Eq. (5.7) it is convenient to write it in a slightly different way. To do so we introduce $e_a = -e$, $e_b = e$, the characteristic frequencies $\omega_i = \frac{|e_i|B}{m_i}$, and use Eq. (3.30b) to get that,

$$\frac{\mathfrak{D}(\mathfrak{I}_i)}{\varphi kT} = -\frac{15}{2\varphi}\left(\frac{n_a}{m_a}a_a^{(1)} + \frac{n_b}{m_b}a_b^{(1)}\right)$$

$$+\frac{6i}{\varphi}\left(-\frac{n_a}{m_a}\omega_a(a_a^{(0)})^2 + \frac{n_b}{m_b}\omega_b(a_b^{(0)})^2\right) + \frac{3i}{\varphi}\left(-\frac{5}{2}\frac{n_a}{m_a}\omega_a(a_a^{(1)})^2 + \frac{5}{2}\frac{n_b}{m_b}\omega_b(a_b^{(1)})^2\right)$$

$$+n_a n_b\Big\{\frac{1}{m_a}(a_a^{(0)})^2 + \frac{3}{m_a}(a_a^{(0)})(a_a^{(1)}) + \frac{13}{4m_a}(a_a^{(1)})^2 + \frac{M_1}{m_b}(a_b^{(0)})^2$$

$$-\frac{3}{\sqrt{m_a m_b}}a_a^{(0)}a_b^{(1)} + \frac{15M_1}{2}(a_b^{(0)})^2 - 2\sqrt{\frac{M_1}{m_a m_b}}(a_a^{(0)})(a_b^{(0)}) + \frac{3\sqrt{M_1}}{\sqrt{m_a m_b}}a_b^{(0)}a_a^{(1)}$$

$$-\frac{27M_1^{\frac{3}{2}}}{2\sqrt{m_a m_b}}a_a^{(1)}a_b^{(1)}\Big\} + \sqrt{2}\frac{n_a^2}{m_a}(a_a^{(1)})^2 + \frac{\sqrt{2M_1}}{m_b}\frac{n_b^2}{m_b}(a_b^{(1)})^2 \tag{5.8}$$

We now take the variation of Eq. (5.8) collect terms in the independent variation of $\delta a_a^{(0)}$, $\delta a_a^{(1)}$ and $\delta a_b^{(1)}$ and set each coefficient equal to zero which leads to a set of three equations with three unknowns, namely,

$$\begin{aligned} &\mathbb{A}_1 a_a^{(0)} - \mathbb{A}_2 a_a^{(1)} - \mathbb{A}_3 a_b^{(1)} = 0 \\ &-\mathbb{B}_1 a_a^{(0)} + \mathbb{B}_2 a_a^{(1)} - \tfrac{9}{2} M_1^2 a_b^{(1)} = 5\tau \\ &-\mathbb{C}_1 a_a^{(0)} - \tfrac{9}{2} M_1 a_a^{(1)} + \mathbb{C}_2 a_b^{(1)} = \tfrac{5\tau}{M_1} \end{aligned} \tag{5.9}$$

when we set $n_a = n_b = \frac{n}{2}$, a fully ionized plasma. Further, $n\tau = \varphi^{-1}$ is taken as the mean free collision time. The coefficients in Eq. (5.9) are given by,

$$\begin{aligned} \mathbb{A}_1 &= 2(1 + M_1^2 - 2M_1 - 6i\omega_a\tau) \\ \mathbb{A}_2 &= -3(1 + M_1) \\ \mathbb{A}_3 &= \tfrac{3}{2} M_1^2(\sqrt{M_1} - 1) \\ \mathbb{B}_1 &= -(1 + M_1) \\ \mathbb{B}_2 &= \tfrac{13}{16} + \tfrac{4}{3}\sqrt{2} - 5i\omega_a\tau \\ \mathbb{C}_1 &= -1 + \tfrac{1}{2} M_1 \\ \mathbb{C}_2 &= 5 + \tfrac{4}{3}\sqrt{\tfrac{2}{M_1}} + 5i\omega_a\tau \end{aligned} \tag{5.10}$$

For the simple case $B = 0$ the resulting set is trivial to solve, at least approximately, by setting all powers of $M_1^n \sim (\frac{1}{1836})^n$, $n \geq 1 \sim 0$ in all additions, namely $1 + M_1^n \simeq 1$. The result is that

$$\begin{aligned} a_a^{(0)} &= 2.94\tau \\ a_a^{(1)} &= 1.96\tau \\ a_b^{(1)} &= \tfrac{0.7265}{M_1}\tau \end{aligned} \tag{5.11}$$

as can be easily checked with a desk computer. The solution for $B \neq 0$ is outlined in Appendix E.

To compute the d'_{is} the procedure is completely similar. The form for the $\mathfrak{D}(\mathfrak{I}_i)$ function only changes in the inhomogeneous term which is now given by $\frac{n_i}{n} f_i^{(0)} \vec{c}_i$. When this is multiplied by $-\tau \vec{c}_i$ and integrated over the velocities after setting

$$\mathfrak{I}_i = \sum_{m=0}^{M} d_i^{(m)} S_{\frac{3}{2}}^{(m)}(c_i)$$

and sum is carried over the two species we get that

$$-n \sum_{i=a}^{b} \frac{n_i}{n} \int f_{i(0)} \vec{c}_i \cdot \vec{c}_i \sum_{m=0}^{M} d_i^{(m)} S_{\frac{3}{2}}^{(m)}(c_i) d\vec{c}_i = 3kT \frac{n_a^2}{m_a} n (1 - M_1 \frac{n_b}{n_a})$$

The resulting algebraic set of equations for $d_i^{(0)}$, $d_i^{(1)}$, $i = a, b$ is identical to the one given by Eq. (5.9) except that the inhomogeneous term changes from

$$\begin{pmatrix} 0 \\ \frac{15}{2}\tau \\ \frac{15}{2M_1}\tau \end{pmatrix} \quad \text{to} \quad \begin{pmatrix} \frac{3}{2}(1 - M_1)\tau \\ 0 \\ 0 \end{pmatrix} \tag{5.12}$$

Once more the solution in the simple case $B = 0$ is readily obtained to read

$$\begin{aligned} d_a^{(0)} &= 1.191\tau \\ d_a^{(1)} &= 0.294\tau \\ d_b^{(1)} &= 0.0234\tau \end{aligned} \tag{5.13}$$

For the case $B \neq 0$ the solution is also outlined in Appendix E.

Bibliography

[1] P. C. Clemmow and J. P. Dougherty; *Electrodynamics of Particles and Plasmas*, Addison-Wesley, Pub-Co., Reading, Mass (1990).

[2] B. B. Robinson and I. B. Bernstein; *Ann Phys*, 110 (1962).

[3] J. O. Hirschfelder, C. F. Curtiss and R. B. Byrd; *The Molecular Theory of Liquids and Gases*, John Wiley & Sons, New York (1964), 2nd ed.

[4] W. Marshall; *The Kinetic Theory of an Ionized Gas*; U.K.A.E.A. Research Group, Atomic Energy Research Establishment. Harwell U.K. parts I, II, and III (1960).

Chapter 6

The Transport Coefficients

In this section we wish to give a detailed account of all the transport coefficients related to the vectorial fluxes discussed in the previous chapters. These are the mass flux $\vec{J}_a(= -\vec{J}_b)$, the corresponding charge flux or electrical current $\vec{J}_c$, closely related to $\vec{J}_a$, and $\vec{J}'_q$ the heat flux. In every case the magnetic field is chosen as the direction of the z-axis, $\vec{B} = B\hat{k}$ so that for any vector, its different components respect to $\vec{B}$ will follow from the decomposition illustrated in Fig. (4.1).

We shall begin our discussion with the heat flux $\vec{J}'_q$ which, according to Eq. (4.10) is given by

$$\vec{J'}_q = -\frac{5}{2}k^2T\sum_i \frac{n_i}{m_i}\Big\{\alpha_i^{(1)}(\mathrm{grad}T)_\parallel + a_i^{(1)(1)}(\mathrm{grad}T)_\perp + a_i^{(2)(1)}B(\mathrm{grad}T)_s +$$

$$(\pm 1)^{ab}T\left[\delta_i^{(1)}(\vec{d}_{ij})_\parallel + d_i^{(1)(1)}(\vec{d}_{ij})_\perp + d_i^{(2)(1)}(\vec{B}\times\vec{d}_{ij})\right]\Big\} \tag{6.1}$$

where in the second term the minus sign has to be taken into account since $\vec{d}_{ij} = -\vec{d}_{ji}$. All the coefficients $\alpha_i^{(1)} = a_i^{(1)(1)} + B^2 a_i^{(3)(1)}$ and $\delta_i^{(1)} = d_i^{(1)(1)} + B^2 d_i^{(3)(1)}$ are the coefficients given in Eqs. (5.11) and (5.13) and Appendix E. Also, the mean free time has been defined as the inverse of the collision integral $[\vec{w}_1, \vec{w}_1]_{12} \equiv \varphi$ in terms of which all collision integrals may be expressed. Thus,

$$n\tau = \frac{1}{\varphi} = \frac{4(2\pi)^{\frac{3}{2}}\sqrt{m_e}\epsilon_0^2(kT)^{\frac{3}{2}}}{e^4\psi} \tag{6.2}$$

Here ϵ_0=8.554 $\times$ 10^{-12} F/m, m_e is the electron mass and ψ is the shielding

L.S. García-Colín, L. Dagdug, *The Kinetic Theory of a Dilute Ionized Plasma*, Springer Series on Atomic, Optical, and Plasma Physics 53

or logarithmic function and is defined as

$$\psi = \ln\left\{1 + \left[\frac{16kTd_0\epsilon_0}{e^2}\right]^2\right\}$$

where d_0 is Debye's length given by,

$$d_0 = \left(\frac{\epsilon_0 kT}{ne^2}\right)^{\frac{1}{2}} \tag{6.3}$$

It follows then from Eq. (6.1) that there are three components for the conventional heat current, one parallel to $\vec{B}$ which is unaffected by the field, one in the (x, y) plane perpendicular to $\vec{B}$ and the third one in the $(-x, y)$ plane whose direction is perpendicular to both $(\vec{J}'_q)_\parallel$ and $(\vec{J}'_q)_\perp$. This is the current known in the literature as the Righi-Leduc effect, the thermal analog of the well known Hall's effect in electromagnetism. But we also have three other contributions to the heat current arising from the diffusive $\vec{d}_{ij}$. In a completely ionized gas, $n_i = n/2$, $\text{grad}(n_i/n) = \vec{0}$. In this case the ordinary or "Fickian" contribution to the heat current does not exist but there will be two other contributions arising from the pressure gradient present in the diffusive force, and another one which includes Thomson's well known thermoelectric effect arising from the term proportional to $\vec{E}'$. These cross terms have been rarely dealt with in the literature. Let us first analyze the three conventional heat currents. Clearly from Eq. (6.1), using the fully ionized plasma condition and noting that $M_1 = \frac{m_e}{m_p} = \frac{1}{1836} < 1$, we have that

$$(\vec{J'}_q)_\parallel = -\kappa_\parallel (\text{grad}T)_\parallel$$

where

$$\kappa_\parallel = \frac{5}{4}\frac{nk^2T}{m_e} 2.01\tau \tag{6.4a}$$

and, in a similar fashion,

$$\kappa_\perp = \frac{5}{4}\frac{nk^2T}{\Delta_1 m_e}\tau(38.7 + 2270x^2 + 161x^4) \tag{6.4b}$$

$$\kappa_s = \frac{5}{4}\frac{nk^2T}{\Delta_1 m_e}\tau(206x + 2644x^3) \tag{6.4c}$$

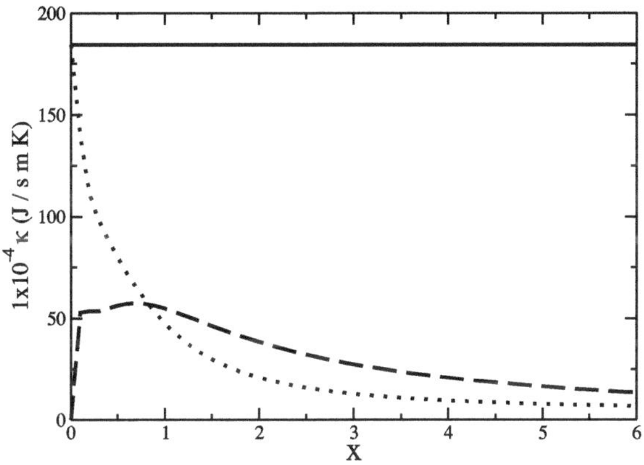

Figure 6.1: The thermal conductivities as function of x for $n = 10^{21}$ cm^{-3} and $T = 10^7$ K. The full line is $\kappa_{\parallel}$, the dotted line is $\kappa_{\perp}$ and the dashed line is κ_s.

where

$$\triangle_1 = 19 + 2078x^2 + 2650x^4 + 9x^6 \tag{6.4d}$$

and $x = \omega_e \tau$, where $\omega_e = \frac{eB}{m_e}$. The three conductivities are plotted in Fig. 6.1 for $0 < x < 6$ and are given in J/s Km. Although the $(kT)^{\frac{5}{2}}$ dependence is the same as in the Spitzer-Braginski calculations, a closer comparison of the results will be given in Chapter 7. Notice also from Eqs. (6.4a-6.4c) that when $x \to 0$ $\kappa_{\perp} \to \kappa_{\parallel}$ as expected, and $\kappa_s \to 0$, and $\vec{J'}_q$ is simply governed by Fourier's equation.

Although in ordinary irreversible thermodynamics a density gradient gives rise to a heat flux, the diffusion thermal (or Dufour) effect, here, if the plasma is fully ionized such effect is not present. Nevertheless the two remaining terms in $\vec{d_{ab}}$ do contribute,

$$\vec{d_{ab}} = \frac{n_a n_b}{n\rho p}(m_a - m_b) \text{ grad } p - \frac{n_a n_b}{p\rho}(m_b e_a - m_a e_b)\vec{E'}$$

the former is a pressure effect and the latter the electromagnetic effect. If we associate species a with the electrons and b with the protons, remembering that $p = nkT$, $n = n_a + n_b$ and $\frac{m_a}{m_b} << 1$, the above equation reduces to,

$$\vec{d_{ab}} = -\frac{1}{2\pi kT} \text{ grad } p - \frac{e}{2kT}\vec{E'} \tag{6.5}$$

Using this result in Eq. (6.1) we find that the three coefficients arising from the pressure gradient are, calling them D,

$$D_\parallel = \frac{5}{4}\frac{n(kT)^2}{m_e} 0.29\tau \tag{6.6a}$$

If we write

$$(\vec{J}_q')_d = D_\parallel (\vec{d_{ij}})_\parallel \tag{6.6b}$$

and

$$(\vec{d_{ij}})_\parallel^{(d)} = -\frac{1}{2}\frac{\text{grad } p}{p}$$

this is convenient since when n =const., $\frac{\text{grad } p}{p} = \frac{\text{grad } T}{T}$, Eq. (6.6b) if combined with Eq. (6.4a) appears to yield an effective thermal conductivity

$$\kappa_\parallel^{ef} = \kappa_\parallel + \frac{1}{2} D_\parallel$$

Taking $\frac{1}{2}\frac{\text{grad } p}{p}$ as the thermodynamic force, the other two coefficients will be given by,

$$D_\perp = \frac{5}{4}\frac{n(kT)^2}{\Delta_1 m_e}\tau(5.6 - 66.2x^2 - 0.22x^4) \tag{6.7a}$$

and

$$D_s = \frac{5}{4}\frac{n(kT)^2}{\Delta_1 m_e}\tau(6.47x + 0.22x^3) \tag{6.7b}$$

The behavior of the coefficients as function of x is shown in Fig. 6.2 whereas the effective conductivities are displayed in the original paper [8]. The pressure contributions do not enhance normal heat flows by more than 7%. Since the force here is $\frac{\text{grad } p}{p} = \frac{\text{grad } p}{nkT}$, these coefficients turn out to be inversely proportional to the density as expected from the general tenets of the kinetic theory of gases.

Finally, the contributions arising from the $\vec{E}'$ term in Eq. (6.1) also deserve some attention. Form Chapter 4 it is clear that

$$\begin{aligned} (\vec{d_{ab}})_\parallel^{(e)} &= E_z \hat{k} \\ (\vec{d_{ab}})_\perp^{(e)} &= (E_x + u_y B)\hat{\imath} + (E_y - u_x B)\hat{\jmath} \\ (\vec{d_{ab}})_s^{(e)} &= -(E_y - u_x B)\hat{\imath} + (E_x + u_y B)\hat{\jmath} \end{aligned} \tag{6.8}$$

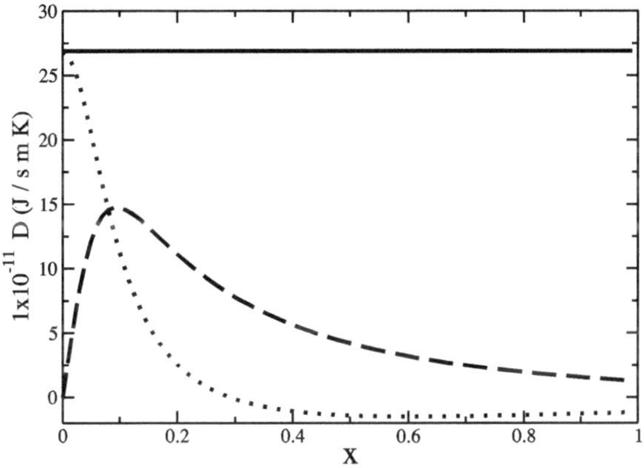

Figure 6.2: The diffusion thermoeffect coefficients as function of x for $n = 10^{21}$ cm^{-3}, $T = 10^7$ K. The full line is $D_{\parallel}$, the dotted line is $D_{\perp}$ and the dashed line is D_s.

so that if $\vec{B} = \vec{0}$ we get the contribution of an electric field $\vec{E} = -e$ grad ϕ to the heat current. This is known as electropyrosis or the Benedicks effect which remains unchanged in the direction of the magnetic field of $\vec{B} \neq \vec{0}$. The other two coefficients should be referred to as electromagnetic pyrosis and their value are given by using the results of appendix E, and denoting them by B,

$$B_{\parallel} = \frac{5}{8} nkT \frac{e}{m_e} 0.29\tau \tag{6.8a}$$

$$B_{\perp} = \frac{5}{8} nkT \frac{e}{m_e} \tau \frac{3.75 - 44.1x^2 - 0.51x^4}{\triangle_1(x)} \tag{6.8b}$$

and

$$B_{\perp} = \frac{5}{8} nkT \frac{e}{m_e} \tau \frac{43.15x + 0.15x^3}{\triangle_1(x)} \tag{6.8c}$$

Their values as functions of x are given in Fig. 6.3. These cross effects, to our knowledge have never been dealt with in the literature. (See however Ref. [9].)

The analysis of the mass and charge transport may be performed simultaneously on account of Eq. (2.15). Indeed the charge current $\vec{J}_c$ is given

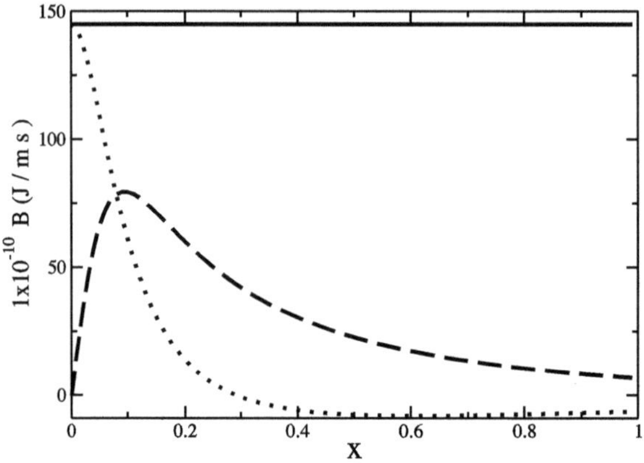

Figure 6.3: The Benedicks coefficients as function of x for $n = 10^{21}$ cm^{-3} and $T = 10^7$ K. $B_{\|}$ is the full line, $B_{\perp}$ is the dotted line and B_s the dashed line.

by

$$\vec{J}_c = \frac{m_a + m_b}{m_a m_b} e\vec{J}_a \simeq \frac{e}{m_a}\vec{J}_a$$

since $\frac{m_a}{m_b} << 1$. Using the definition of $\vec{J}_a$,

$$\vec{J}_c = \frac{e}{m_a}\int \vec{c}_a f_a^{(0)} \varphi_a^{(1)} d\vec{c}_a$$

with $\varphi_a^{(1)}$ defined in Eq. (3.13). Using Eq. (4.2) we then have that

$$\vec{J}_c = \frac{n_a e}{m_a} k \left[\alpha_a^{(0)} (\text{grad } T)_{\|} + \mathrm{a}_a^{(1)(0)} (\text{grad } T)_{\perp} + \mathrm{a}_a^{(2)(0)} B (\text{grad } T)_s\right] +$$

$$\frac{n_a e}{m_a} kT \left[\delta_a^{(0)} (\vec{d}_{ab})_{\|} + d_a^{(1)(0)} (\vec{d}_{ab})_{\perp} + d_a^{(2)(0)} (d_a B\hat{k} \times \vec{d}_{ab})\right] \qquad (6.9)$$

The first three terms on the right hand side contain the contribution of a thermal gradient to an electric current, the well known thermoelectric effect whereas the remaining three terms contain the pressure gradient effects on the electric current plus the ordinary electrical conductivity when $\vec{B} = 0$ and the electromagnetic contribution to the electro magnetic contribution to the electric current when $\vec{B} \neq \vec{0}$. We shall discus the first and third one separately and will come back to the second one later on. Let begin with the first three terms. If $\vec{B} = \vec{0}$, $(\text{grad } T)_\perp + (\text{grad } T)_\| =$ grad T, $\text{a}_a^{(1)(0)} = d_a^{(1)(0)}$, and

$$\vec{J}_c = -\tau_k \text{ grad } T$$

where τ_k is the standard Thomson thermoelectric coefficient given by

$$\tau_k = \frac{2.94 n_a k e}{m_e} \tau \tag{6.10a}$$

where τ has already been defined. If $\vec{B} \neq \vec{0}$

$$(\tau_k)_\perp = \frac{5 n_a k e \tau}{\triangle_1(x) m_e}(7.54 + 430x^2 + 1.46x^4) \tag{6.10b}$$

and

$$(\tau_k)_s = \frac{5 n_a k e \tau}{\triangle_1(x) m_e}(41.22x + 530x^3 + 18x^5) \tag{6.10c}$$

where $\triangle_1(x)$ is defined in Eq. (6.4d). The three Thomson coefficients are depicted in Fig. 6.4. Clearly. $(\tau_k)_\perp \to (\tau_k)_\|$ when $B \to 0$ whereas $(\tau_k)_s \to 0$. Nevertheless, this is a common feature of all transport coefficients, there is a range of values of $x = \omega_e \tau$ where the perpendicular and Righi-Leduc like contributions to the full thermoelectric effect are not negligible.

We now turn to the electrical conductivity in the z-direction, if $E_z \neq 0$ is not modified by the magnetic field, so that from Eq. (6.9) we get that

$$\sigma_\| = \frac{n_a e^2}{2 m_a} 1.191 \tau \tag{6.11a}$$

which goes as $(kT)^{\frac{3}{2}}$ as predicted by other authors. On the other hand if $\vec{B} \neq \vec{0}$

$$\sigma_\| = \frac{3 n_a e^2 \tau}{4 \triangle_1 m_a}(15.21 + 36.83x^2 + 0.1224x^4) \tag{6.11b}$$

and

$$\sigma_\| = \frac{3 n_a e^2 \tau}{4 \triangle_1 m_a}(156x + 221.2x^3) \tag{6.11c}$$

The three conductivities are shown in Fig 6.5 as function of x. As said above the mass fluxes are identical to the charge fluxes if these are multiplied by $1/e$.

Since for a fully ionized plasma there is no Fickian diffusion $\nabla(n_i/n) = 0$ the transport coefficients arising from the pressure tensor are proportional

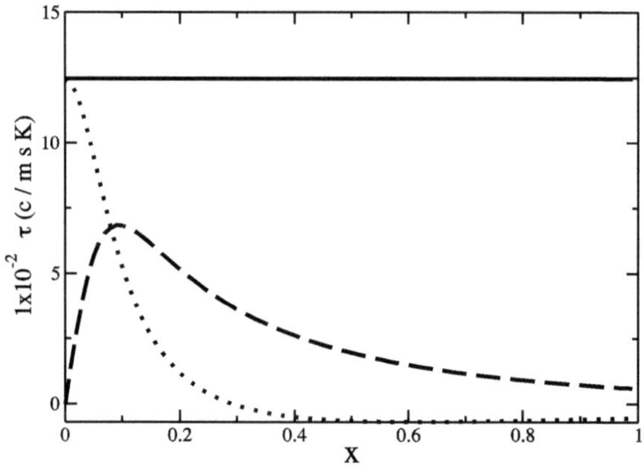

Figure 6.4: The three thermoelectric or Thomson coefficients shown as function of x for $n = 10^{21}$ cm^{-3} and $T = 10^7$ K. The full line is the parallel coefficient $\tau_{\|}$, the dotted line is $\tau_{\perp}$ and the dashed line is the "Righi-Leduc" type coefficient τ_s.

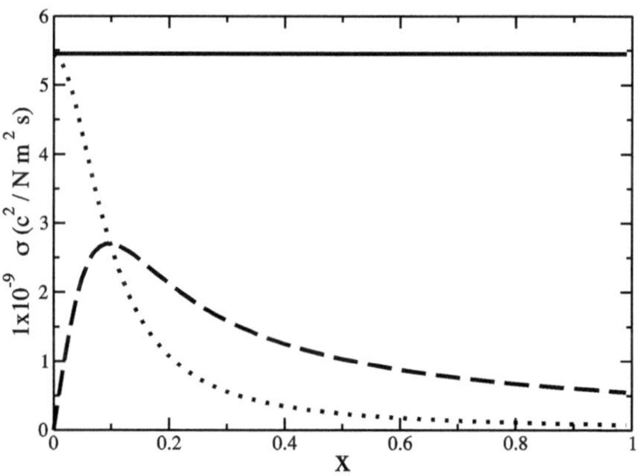

Figure 6.5: The three electrical conductivities plotted as function of x for $n = 10^{21}$ cm^{-3} and $T = 10^7$ K. The full line is $\sigma_{\|}$, the dotted line is $\sigma_{\perp}$ and the dashed line σ_s.

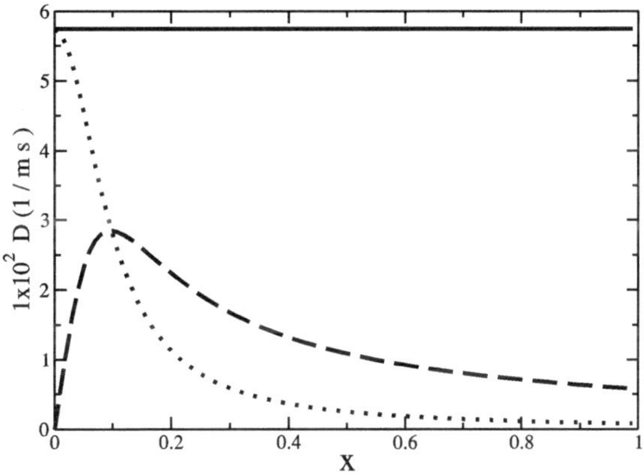

Figure 6.6: The diffusion conductivities plotted as function of x for or $n = 10^{21}$ cm^{-3} and $T = 10^7$ K. The full line is $D_{\|}$, the dotted curve is $D_{\perp}$, and the dashed line corresponds to D_s.

to those appearing in $\vec{J}_c$ which we have not written down but are trivial to obtain. Since they apparently play no important role, in practice we shall not bother with them. The remaining contributions to $\vec{J}_a$ namely, those that come from $\vec{d}_{ab}^{(e)}$ are essentially those written in Eqs. (6.11a-6.11c) which when $\vec{B} = \vec{0}$ are known in the literature as the "Dorn" effect. On the other hand the effects of a thermal gradient as mass diffusion or the Soret effect arise from the first three terms in Eq. (6.9) when multiplied by m_a/e. There is no need to write their mathematical expressions since they are similar to the Thomson coefficients, but for the sake of completeness we have plotted them as function of x in Fig. 6.6.

When discussing the direct diffusion process in a fully ionized plasma, we emphasized that true Fickian diffusion namely the contribution of a concentration gradient to a mass flux $\vec{J}_a(= -\vec{J}_b)$ vanishes. This of course is no longer true in a partially ionized plasma for which $n_a = xn_b$, $n_b = (1+x)^{-1}n$, $x(\vec{r}, t)$. In this case, the variational procedure leading to the results of Appendix C has to be repeated setting $\frac{n_a}{n_b} = x$ and the same modification will apply to the coefficient of grad $x(1+x)^{-1}$ as well as the diffusive contribution to $\vec{J}_q'$. This will gives rise to the true diffusion thermo effect, the originally named Dufour effect. Since this calculation is straightforward we shall leave it as an exercise for the interested reader.

With all the results given in this chapter it is now clear that the complete structure of the constitutive equations required to write up the equations of magnetohydrodynamics is far more subtle than often quoted in the literature since their structure will be strongly dependent on the values of n, T, and B. We shall fully deal with this equation in the last chapter of the book.

Bibliography

[1] L. Spitzer, R. S. Cohen and P. Mc Roictly; *Phys. Rev.* **80**, 230 (1950).

[2] L. Spitzer and R. Härm; *Phys. Rev.* **89**, 977 (1953).

[3] L. Spitzer; *The Physics of Full Ionized Gases*, Wiley-Interscience Publ. Co., New York (1962).

[4] S. I. Braginskii; *Transport Processes in a Plasma*; in Plasma Physics Reviews Consultants Bureau Enterprise, N.Y. 1965.

[5] F. L. Hinton, R. D. Hazeltime; *Rev. Mod. Phys.*, **48**, 239 (1976).

[6] R. Balescu; *Transport Processes in Plasmas*; Vol. I. Classical Transport, North-Holland Publ. Co., Amsterdam (1988).

[7] J. M. Burgers; *Flow Equations for Composite Gases*; Academic Press, Inc., New York (1969).

[8] L. S. García-Colín, A. L. García-Preciante and A. Sandoval-Villalbazo; *J. Phys. Plasmas* **14**, 012305 (2007); ibid **14**, 08990 (2007).

[9] L. S. García-Colín, A. L. García-Preciante and A. Sandoval-Villalbazo; *J. Non-equilib. Thermodyn.* **32**, 379 (2007)

[10] See Ref. [5] Chapter 4.

Chapter 7

Discussion of the Results

The results obtained along the several sections of this work require an objective analysis to place them within the actual tendencies in the field of the kinetic theory of plasmas. They can be divided into two classes, one related to the general structure of magnetohydrodynamics and the other one, containing directly the essential information displayed in the different transport coefficients. In both cases we shall take to be as a reference point the magnificent and exhaustive treatment, very likely the most complete one written on this subject, due to R. Balescu and published already 18 years ago [1].

With respect to magnetohydrodynamics many important features of the subject were brought up by Balescu in his book [1]. Yet for the sake of completeness and to fully account for all missing details we shall bring up the subject in Chap. 9. Just to give a qualitative introduction to the problem, it is worth mentioning that Balescu did not took into account the cross coefficients which appear in the mass (or charge) balance equations nor in the heat vector, the energy (or temperature) transport equation. Thus for instance in Eq. (2.24), when $\vec{J'}_q$ as given in (4.10) is introduced, besides the three contributions arising from the heat conduction terms,

$$\frac{5}{2}k^2T\sum_{i=a}^{b}\frac{n_i}{m_i}\ \text{div}\ \left[\mathrm{a}_i^{(1)(1)}\ \text{grad}\ T+\mathrm{a}_i^{(2)(1)}\vec{B}\times\ \text{grad}\ T+\mathrm{a}_i^{(2)(1)}\vec{B}^2\ \text{grad}\ T\right]$$

there are three more terms coming from the diffusive term. If $\vec{B}$ turns out to be a constant magnetic field the div operation will not give rise to anything involving explicitly $\vec{B}$ but if $\vec{B}$ is inhomogeneous,

$$\text{div}\ (\vec{B}\times\ \text{grad}\ T)=\ \text{grad}\ T\cdot\ \text{rot}\ \vec{B}$$

L.S. García-Colín, L. Dagdug, *The Kinetic Theory of a Dilute Ionized Plasma*, Springer Series on Atomic, Optical, and Plasma Physics 53

and

$$\text{div } (B^2 \text{ grad } T) = \text{ grad } T \cdot \text{ grad } B^2 + B^2 \nabla^2 T$$

so that

$$\text{div } (\vec{J'}_q)_{elec} = \frac{5}{2} k^2 T \sum_{i=a}^{b} \frac{n_i}{m_i} \Big\{ \text{a}_i^{(1)(1)} \nabla^2 T + \text{a}_i^{(2)(1)} \text{ grad } T \cdot \text{ rot } \vec{B} +$$

$$\text{a}_i^{(3)(1)} (B^2 \nabla^2 T + (\text{ grad } B^2) \cdot (\text{ grad } T)) \Big\} \tag{7.1}$$

which considerably complicates Eq. (2.24). Three similar contributions arise from the diffusive terms in Eq. (4.8) which we shall not write down explicitly. From Eq. (7.1) we see that in this case, $\vec{B}$ has to be explicitly obtained from Maxwell's equations and further, its contribution to the energy transport is far more complicated than if we use the simpler constitutive equation

$$\vec{J'}_q = -\kappa_{\|} (\text{ grad } T)_{\|} - \kappa_{\perp} (\text{ grad } T)_{\perp} - \kappa_s (\text{ grad } T)_s \tag{7.2}$$

which is strictly valid for constant magnetic fields.

Completely analogous comments follow for either mass or charge transport as depicted in Eq. (4.2). If $\vec{B}$ is inhomogeneous, div $\vec{J}_a$ has to be calculated directly from Eq. (4.1) so that both the thermal contributions to heat transport have a structure similar to Eq. (7.1) and the same can be said about the magnetic contributions to mass transport arising from the second line of (4.1). Writing all these equations is rather unnecessary unless explicitly required in a very particular situation. The interested reader may do it with a minimum effort if desired. Finally a word about the term $\overleftrightarrow{\tau}^k$: grad $\vec{u}$. The full stress tensor $\overleftrightarrow{\tau}^k$ in the presence of a magnetic field is determined completely in Chap. 8 together with the values of the five independent viscosities so that the different contributions arising from this term are also determined but rather needless writing them explicitly. We leave that as an exercise for the reader.

The main issue then is that the complete structure of the magnetohydrodynamic equations for a dilute, inert binary mixture of charged particles, is here accomplished within the limitations imposed by the condition that $\omega_i \tau_i \sim 1$. We might remind the reader that this condition has to be read very carefully since although for electrons $\omega_e > \omega_p$ and $\omega_e \sim B$, τ_i is both a function of the temperature and the density so it may happen that not so small fields can match up with temperature and density values such that the criterion is met. For the values of these latter variables characteristic of many astrophysical systems B turns out to be very small ($\sim 10^{-6}\mu$G) so modifications have to be made to account for fields at least 10^{-6} G higher. As mentioned, magnetohydrodynamics will be exhaustively discussed in Chap. 9. In the case of neo-classical transport the results obtained by this method are certainly not valid [2] but these situations are completely outside of the scope of this work.

Coming back to the second class of results, the values of the transport coefficients obtained here for self consistent magnetic fields, it is convenient to compare them with those that have been obtained in the literature by other methods. The most relevant calculations along these lines are the Spitzer-Braginski's ones (see ref. (1)-(4) Chapter 6) and those obtained by Balescu [2]. Since in his book the analysis of the Spitzer-Braginski's results is made in full detail we refer the reader to that source and limit our discussion to a comparison with those of Balescu. Before going into the details, it is important to keep in mind that this author uses the Landau form of the Boltzmann collision kernel, assumes that the electrons and ions are independent in local equilibrium which is not that important and thirdly, he solves the resulting kinetic equation using a Grad like moment expansion. It is this last feature of his work which plays a very important role in evaluating the final form for the transport coefficients. The moment method which originated with H. Grad in kinetic theory [3] has been used in various forms and with several purposes in deriving transport equations. We shall leave this particular problem aside. Also, it served to attempt giving a kinetic basis to what was referred to in the literature as the kinetic basis of Extended Irreversible Thermodynamics and in that context used to discuss many aspects of transport theory [4], [5]. A few years ago one of us in collaboration with other people [6] pointed out that the method may lead to inconsistencies when used to obtain transport equations and that the method used to compute transport coefficients leads to ambiguities insofar as knowing precisely what different orders in the gradients contribute to the value of anyone of them. Further, it was shown that [7] this question is inevitable unless an ordering parameter is brought in to the scheme to separate the different contributions to different orders in the gradients. This difficulty with the moment method had already been pointed out by Grad himself in his seminal contribution to the subject [3]. This is the enormous advantage of Knudsen's parameter in the Chapman-Enskog expansion. Thus the discrepancies that we get with Balescu's results may be due to this facet of the method.

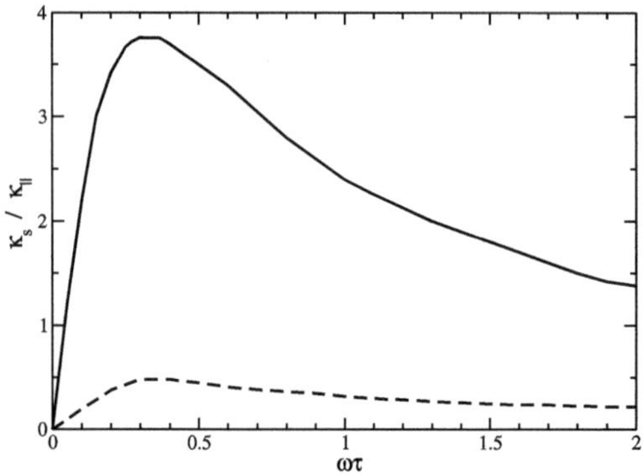

Figure 7.1: The adimensional factors for the parallel and Righi-Leduc conductivities as defined by Balescu (Ref. [2]) plotted against x. The dashed line is Balescu's result, the full line is our result. The difference is explained in the text.

When we compare our results with his we take his values for $Z = 1$ and translate our graphs to his language namely use the CGS system and most important, plot the values vs. $x = \omega_e \tau_e$ as we did in Chap. 6. In Fig. (7.1) we have the representative aspect of this comparison. The parallel conductivities are essentially the same but this is not so for the other two. The small differences may be explained since our result, we insist are valid only to first order in the gradients; his results come from the 21 moments representation and it is not clear what orders in the gradients are here involved [7]. Similar argument is obtained for the electrical conductivity, both methods yield qualitatively the same values.

As for the cross coefficients we refer the reader to an exhaustive paper on the subject which has recently been published [10]. One more comment is desirable. Since W. Marshall did pioneering work on this problem following essentially the same approach taken here, in fact, we were highly motivated by his reports, it is very pertinent to fully discuss the differences in the results. To accomplish this comparison let us explain in detail where our results begin to differ. For this purpose, we recall report III of his work, the relevant one, here and will indicate the equations in his work with a capital M, see Appendix I. We begin with the equation for the electrical current written at the bottom of p. 22 leading to equation M3.6.1. This result is identical to the one quoted in our text, Eq. (3.10). Next he defines a new

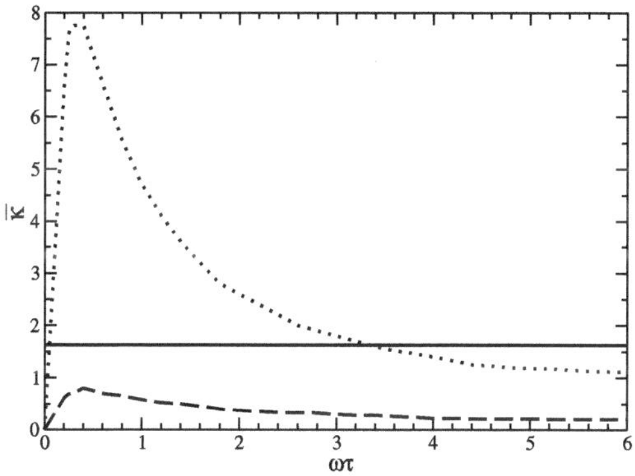

Figure 7.2: The ratio $\kappa_s/\kappa_{\|}$ as calculated by our method (short dashed line) compared with Marshall's (long dashed line). The solid line is just a reference value taken as $\kappa_{\|}$.

diffusive force $\vec{D}$ given by

$$\vec{D} = -\frac{p\rho}{e_1 m_2 - e_2 m_1}\vec{d_1}$$

where $\vec{d_1}$ is given in (M3.14). Then he writes the electrical current in (M3.62) and concludes that the coefficients σ_I, σ_{II} and σ_{III} are the thermal diffusion coefficients. This is wrong. First of all, the force $\vec{D}$ has three components, a coefficient times $\nabla(\frac{1}{n_2})$, a coefficient times ∇p and the coefficient of $\vec{E}'$ (see M3.63). The former one, a concentration gradient that produces an electrical current is called the "Dorn effect". The coefficients of $\vec{E}'$ are the electrical conductivities when $\nabla p = \vec{0}$, the pressure contribution to $\vec{J_c}$. The coefficients ϕ^I, ϕ^{II} and ϕ^{III}, those arising from thermal gradient contributing to $\vec{J_c}$ are the thermo electric or Thomson's coefficients none of which have anything to do with thermal diffusion.

As pointed out before, mass and electric currents are easily related to each other by for the numerical factor $(m_1 + m_2)^{-1} m_1 m_2 e$. This implies that the ϕ^{I-III} coefficients are indeed the same as the three Soret coefficients which we have exhibited in Chapter 6. On the other hand, except for a numerical scaling factor the coefficients σ_I, σ_{II} and σ_{III} (Eqs. M7.8 and M7.10) are proportional to the ordinary diffusion coefficients not present in a fully ionized plasma since $n_a = n_b = n/2$.

Next, we consider the equation for the flow of heat. In the form written by Marshall, in Eq. (M7.16) the θ^{I}, θ^{II} and θ^{III}, the thermal conductivities as defined in Eq. (M7.19) are not precisely $\kappa_{\|}$, $\kappa_{\perp}$ and κ_{s} because our definitions of heat flow differ by the term $\frac{5}{2}kT(n_1 < \vec{c}_1 > + n_2 < \vec{c}_2 >)$. Indeed according to classical irreversible thermodynamics [8] this term must be included to get $\vec{J}'_q$ as done in section 4.2 of this book. Marshall on the other hand deals with it a la old fashioned form of Chapman and Cowling leading to complicated expressions for the thermal conductivities. Nevertheless the θ' and κ' have been here compared and the results shown in Fig. (7.2), see Ref. [10]. It is important to emphasize that the expression here used as definition of the heat flux allows a clear separation between thermodynamic fluxes and forces as has become standard since the classical work of de Groot and Mazur [8]. Thus we may assert that what we achieve in this book is a rather comprehensive derivation of the transport processes of a dilute inert plasma within the tenets of irreversible thermodynamics.

Bibliography

[1] See Ref. [6] in Chap 5.

[2] R. Balescu; *Transport Processes in Plasmas; Vol. I. Classical Transport*; North-Holland Publ. Co., Amsterdam (1988).

[3] H. Grad; *Principles of the Kinetic Theory of Gases*; in Handbuch der Physik, S. Flügge, Ed., Vol. 12; Springer-Verlag, Berlin (1958).

[4] L. S. García-Colín and R. M. Velasco; *J. Nonequilib. Thermodyn* **18**, 157 (1993) and Ref. cited therein.

[5] R. M. Velasco and L. S. García-Colín; *J. Stat. Phys.*, **69**, 217 (1992).

[6] R. M. Velasco, F. J. Uribe and L. S. García-Colín; *Phys Rev. E*, **66**, 32103 (2002).

[7] L. S. García-Colín, R. M. Velasco and F. J. Uribe; *J. Nonequilib. Thermodyn.* **29**, 257 (2004).

[8] S. R. de Groot and P. Mazur; *Non-equilibrium Thermodynamics*; Dover Publications, Inc., Mineola, N.Y. (1984).

[9] See Ref. [5] Chap. 4.

[10] L. S. García-Colín, A. L. García-Perciante and A. Sandoval-Villalbazo; *Phys. Plasmas* **14**, 012305 (2007); ibid **14**, 089901 (2007).

Part II

Tensorial Transport Processes

Chapter 8

Viscomagnetism

8.1 The Integral Equation

In this section we wish to consider the contribution of the term $\overleftrightarrow{\mathbb{B}} : \text{grad}\vec{u}$ to the transport processes occurring in a magnetized dilute plasma. For this purpose we must seek the most general solution to Eq. (3.9) which reads as:

$$\varphi_i^{(1)} = \overleftrightarrow{\mathbb{B}}_i : \text{grad } \vec{u} + \vec{\mathbb{A}}_i \cdot \text{grad } \ln T + \overleftrightarrow{\mathbb{D}}_i \cdot \vec{d_{ij}} \tag{3.13}$$

where on account of the inhomogeneous term in that equation, $\overleftrightarrow{\mathbb{B}}_i$ has to be the most general symmetric traceless tensor which may be constructed from the vectors $\vec{c_i}$, $\vec{c_i} \times \vec{B}$, and $(\vec{c_i} \times \vec{B}) \times \vec{B} = \vec{B}(\vec{c_i} \cdot \vec{B})$. Clearly all scalars appearing in such a tensor will be function of c_i^2, B^2, $(\vec{c_i} \cdot \vec{B})^2$, n, T, etc. When Eq. (3.13) is substituted back into Eq. (3.9) we get that

$$\frac{m_i}{kT}\overleftrightarrow{c_i{}^0 c_i} = \frac{e_i}{m_i}(\vec{c_i} \times \vec{B}) \cdot \frac{\partial}{\partial \vec{c_i}} \overleftrightarrow{\mathbb{B}}_i -$$

$$\frac{m_i}{\rho k T}\left(\sum_j \left(e_j \left[\int d\vec{c_j} f_j^{(0)} \vec{c_j}\right] \times \vec{B}\right) \overleftrightarrow{\mathbb{B}}_j\right) \cdot \vec{c_i} + C(\overleftrightarrow{\mathbb{B}}_i) + C(\overleftrightarrow{\mathbb{B}}_i + \overleftrightarrow{\mathbb{B}}_j) \tag{8.1}$$

$$\text{for } \; i, j = a, b$$

Besides the unit tensor $\mathbb{I}$ only six symmetric traceless tensors may be constructed from the three independent vectors and by inspection they readily

L.S. García-Colín, L. Dagdug, *The Kinetic Theory of a Dilute Ionized Plasma*,
Springer Series on Atomic, Optical, and Plasma Physics 53

follow,

$$\overleftrightarrow{\mathcal{T}}_i^{(0)} = \mathbb{I}$$

$$\overleftrightarrow{\mathcal{T}}_i^{(1)} = \overleftrightarrow{c_i{}^0 c_i}$$

$$(\overleftrightarrow{\mathcal{T}}_i^{(2)})_{\alpha\beta} = \tfrac{1}{2}\left[\vec{c}_{i\alpha}(\vec{c}_i \times \vec{B})_\beta + \vec{c}_{i\beta}(\vec{c}_i \times \vec{B})_\alpha\right]$$

$$(\overleftrightarrow{\mathcal{T}}_i^{(3)})_{\alpha\beta} = (\vec{c}_i \times \vec{B})_\alpha(\vec{c}_i \times \vec{B})_\beta - \tfrac{1}{3}\left(B^2 c_i^2 - (\vec{B}\cdot\vec{c}_i)^2\right)\delta_{\alpha\beta} \tag{8.2}$$

$$(\overleftrightarrow{\mathcal{T}}_i^{(4)})_{\alpha\beta} = \tfrac{1}{2}\left[\vec{c}_{i\alpha}\vec{B}_\beta + \vec{c}_{i\beta}\vec{B}_\alpha\right](\vec{c}_i \cdot \vec{B}) - \tfrac{1}{3}(\vec{c}_i \cdot \vec{B})^2\delta_{\alpha\beta}$$

$$(\overleftrightarrow{\mathcal{T}}_i^{(5)})_{\alpha\beta} = \tfrac{1}{2}\left[\vec{B}_\alpha(\vec{c}_i \times \vec{B})_\beta + \vec{B}_\beta(\vec{c}_i \times \vec{B})_\alpha\right](\vec{c}_i \cdot \vec{B})$$

$$(\overleftrightarrow{\mathcal{T}}_i^{(6)})_{\alpha\beta} = \vec{B}_\alpha\vec{B}_\beta(\vec{c}_i \cdot \vec{B})^2 - \tfrac{1}{3}B^2(\vec{c}_i \cdot \vec{B})^2\delta_{\alpha\beta}$$

We now propose following the same technique as in the vectorial processes, that

$$\overleftrightarrow{\mathbb{B}}_i = \sum_{n=0}^{6} \Gamma_i^{(n)} \overleftrightarrow{\mathcal{T}}_j^{(n)} \qquad \text{for } i = a, b \tag{8.3}$$

where Γ_i is a function of all scalars in the field. Notice that for all n, $\overleftrightarrow{\mathcal{T}}_i^{(n)}$ is an even function of the velocities $\vec{c}_i$ so upon substitution of Eq. (8.3) into Eq. (8.1) the integral

$$\int d\vec{c}_j f_j^{(0)} \overleftrightarrow{\mathcal{T}}_i^{(n)} (\vec{c}_j \times \vec{B}) = 0 \qquad \text{for all } n$$

since its integrand is an odd function of $\vec{c}_j$, whence Eq. (8.1) reduces to

$$\frac{m_i}{kT}\overleftrightarrow{c_i{}^0 c_i} = \frac{e_i}{m_i}(\vec{c}_i \times \vec{B}) \cdot \frac{\partial}{\partial \vec{c}_i} \sum_{n=0}^{6} \Gamma_i^{(n)} \overleftrightarrow{\mathcal{T}}_i^{(n)} + \text{ collision terms} \tag{8.4}$$

Moreover,

$$(\vec{c}_i \times \vec{B}) \cdot \frac{\partial}{\partial \vec{c}_i} \Gamma_i^{(n)} \overleftrightarrow{\mathcal{T}}_i^{(n)} = (\vec{c}_i \times \vec{B}) \cdot \left(\frac{\partial \Gamma_i^{(n)}}{\partial \vec{c}_i}\right) \overleftrightarrow{\mathcal{T}}_i^{(n)} + (\vec{c}_i \times \vec{B}) \cdot \Gamma_i^{(n)} \left(\frac{\partial \overleftrightarrow{\mathcal{T}}_i^{(n)}}{\partial \vec{c}_i}\right)$$

for all n. But $\frac{\partial \Gamma_i^{(n)}}{\partial \vec{c}_i} = \frac{\vec{c}_i}{|\vec{c}_i|}\frac{\partial \Gamma_i^{(n)}}{\partial \vec{c}_i}$ since $\Gamma_i^{(n)}$ is a scalar function of $\vec{c}_i$ whence, the first term always vanishes, $(\vec{c}_i \times \vec{B}) \cdot \vec{c}_i = 0$. Thus we need to evaluate only the action of the operator $\text{grad}_{c_i} \equiv \frac{\partial}{\partial \vec{c}_i}$ on each tensor $\overleftrightarrow{\mathcal{T}}_i^{(n)}$. Clearly, $\frac{\partial}{\partial \vec{c}_i}\overleftrightarrow{\mathcal{T}}_i^{(0)} = \vec{0}$ and the rest of the terms are given by,

$$(\vec{c}_i \times \vec{B}) \cdot \frac{\partial}{\partial \vec{c}_i}\overleftrightarrow{\mathcal{T}}_i^{(1)} = 2\overleftrightarrow{\mathcal{T}}_i^{(2)} \tag{8.5a}$$

$$(\vec{c}_i \times \vec{B}) \cdot \frac{\partial}{\partial \vec{c}_i}\overleftrightarrow{\mathcal{T}}_i^{(2)} = -B^2\overleftrightarrow{\mathcal{T}}_i^{(1)} + \overleftrightarrow{\mathcal{T}}_i^{(3)} + \overleftrightarrow{\mathcal{T}}_i^{(4)} \tag{8.5b}$$

$$(\vec{c}_i \times \vec{B}) \cdot \frac{\partial}{\partial \vec{c}_i}\overleftrightarrow{\mathcal{T}}_i^{(3)} = -2B^2\overleftrightarrow{\mathcal{T}}_i^{(2)} + 2\overleftrightarrow{\mathcal{T}}_i^{(5)} \tag{8.5c}$$

$$(\vec{c}_i \times \vec{B}) \cdot \frac{\partial}{\partial \vec{c}_i}\overleftrightarrow{\mathcal{T}}_i^{(4)} = \overleftrightarrow{\mathcal{T}}_i^{(5)} \tag{8.5d}$$

$$(\vec{c}_i \times \vec{B}) \cdot \frac{\partial}{\partial \vec{c}_i}\overleftrightarrow{\mathcal{T}}_i^{(5)} = -B^2\overleftrightarrow{\mathcal{T}}_i^{(4)} + \overleftrightarrow{\mathcal{T}}_i^{(6)} \tag{8.5e}$$

$$(\vec{c}_i \times \vec{B}) \cdot \frac{\partial}{\partial \vec{c}_i}\overleftrightarrow{\mathcal{T}}_i^{(6)} = 0 \tag{8.5f}$$

Eqs. (8.5a), (8.5d) and (8.5f) follow almost by inspection but Eqs. (8.5b), (8.5c) and (8.5e) require some lengthy and abnoxious algebra. Probably the easiest way of proving them is by components $\alpha\beta$ and after some hard work the equations prove to be correct. These equations show that the space spanned by the seven tensors of Eq. (8.2) form a closed space, acting by the operator grad_{c_i} does not create new tensors. Nevertheless Eqs. (8.5) are rather inappropriate to manipulate Eq. (8.4) since the inhomogeneous term contains $\overleftrightarrow{c_i{}^0 c_i}$ which does not appear in the base defined by Eqs. (8.2). Therefore we must perform a transformation to another base in which only the symmetric traceless tensors $\overleftrightarrow{c_i{}^0 c_i}$ and $\overleftrightarrow{B\,{}^0 B}$ appear as well as the constant tensors $\mathbb{I}$ and ϵ_{ijk} the Levi-Civita tensor. Eight symmetric tensors may be constructed from these basic ones, namely,

$$
\begin{aligned}
(\overleftrightarrow{Q}_i^{(0)})_{i,\alpha\beta} &= \delta_{\alpha\beta} \\
\overleftrightarrow{Q}_{i,\alpha\beta}^{(1)} &= \delta_{\alpha\gamma}\delta_{\beta\lambda}(\overleftrightarrow{c_i{}^0c_i})_{\gamma\lambda} \\
\overleftrightarrow{Q}_{i,\alpha\beta}^{(2)} &= \tfrac{1}{2}(\delta_{\alpha\gamma}\epsilon_{\beta\lambda\varphi} + \delta_{\beta\gamma}\epsilon_{\alpha\lambda\varphi})B_\varphi(\overleftrightarrow{c_i{}^0c_i})_{\gamma\lambda} \\
\overleftrightarrow{Q}_{i,\alpha\beta}^{(3)} &= \epsilon_{\alpha\gamma\varphi}\epsilon_{\beta\lambda\psi}(\overleftrightarrow{B{}^0B})_{\varphi\psi}(\overleftrightarrow{c_i{}^0c_i})_{\gamma\lambda} \\
\overleftrightarrow{Q}_{i,\alpha\beta}^{(4)} &= \delta_{\alpha\beta}(\overleftrightarrow{B{}^0B})_{\gamma\lambda}(\overleftrightarrow{c_i{}^0c_i})_{\gamma\lambda} \\
\overleftrightarrow{Q}_{i,\alpha\beta}^{(5)} &= c_i^2(\overleftrightarrow{B{}^0B})_{\alpha\beta} \\
\overleftrightarrow{Q}_{i,\alpha\beta}^{(6)} &= \tfrac{1}{2}\left[\delta_{\alpha\gamma}(\overleftrightarrow{B{}^0B})_{\beta\lambda} + \delta_{\beta\gamma}(\overleftrightarrow{B{}^0B})_{\alpha\lambda}\right](\overleftrightarrow{c_i{}^0c_i})_{\gamma\lambda} \\
\overleftrightarrow{Q}_{i,\alpha\beta}^{(7)} &= \tfrac{1}{2}(\epsilon_{\beta\gamma\varphi}(\overleftrightarrow{B{}^0B})_{\alpha\varphi} + \epsilon_{\alpha\gamma\varphi}(\overleftrightarrow{B{}^0B})_{\beta\varphi})B_\lambda(\overleftrightarrow{c_i{}^0c_i})_{\gamma\lambda} \\
\overleftrightarrow{Q}_{i,\alpha\beta}^{(8)} &= (\overleftrightarrow{B{}^0B})_{\alpha\beta}(\overleftrightarrow{B{}^0B})_{\alpha\gamma}(\overleftrightarrow{c_i{}^0c_i})_{\gamma\lambda}
\end{aligned}
\tag{8.6}
$$

The construction of these tensors follow from certain rules on isotropic cartesian tensors too involved to explain here. Also, in Appendix F we outline the proof that the tensors $\overleftrightarrow{\mathcal{T}}$ given in Eq. (8.2) are related to the $\overleftrightarrow{Q}$ tensors as follows,

$$
\begin{aligned}
\overleftrightarrow{\mathcal{T}_i}^{(0)} &= \overleftrightarrow{Q}_i^{(0)} \\
\overleftrightarrow{\mathcal{T}_i}^{(1)} &= \overleftrightarrow{Q}_i^{(1)} \\
\overleftrightarrow{\mathcal{T}_i}^{(2)} &= \overleftrightarrow{Q}_i^{(2)} \\
\overleftrightarrow{\mathcal{T}_i}^{(5)} &= \overleftrightarrow{Q}_i^{(7)} \\
\overleftrightarrow{\mathcal{T}_i}^{(3)} &= \overleftrightarrow{Q}_i^{(3)} + \tfrac{1}{3}\overleftrightarrow{Q}_i^{(4)} - \tfrac{1}{3}B^2\overleftrightarrow{Q}_i^{(1)} - \tfrac{1}{3}\overleftrightarrow{Q}_i^{(5)} \\
\overleftrightarrow{\mathcal{T}_i}^{(4)} &= \overleftrightarrow{Q}_i^{(6)} - \tfrac{1}{3}\overleftrightarrow{Q}_i^{(4)} - \tfrac{1}{3}B^2\overleftrightarrow{Q}_i^{(1)} + \tfrac{1}{3}\overleftrightarrow{Q}_i^{(5)} \\
\overleftrightarrow{\mathcal{T}_i}^{(4)} &= \overleftrightarrow{Q}_i^{(6)} + \tfrac{1}{3}B^2\overleftrightarrow{Q}_i^{(5)}
\end{aligned}
\tag{8.7}
$$

The first and last lines follow at once, but the two middle expressions require a little bit of work. Notice however that, $\overleftrightarrow{\tau_i}^{(3)} + \overleftrightarrow{\tau_i}^{(4)} = \overleftrightarrow{Q}_i^{(3)} + \overleftrightarrow{Q}_i^{(6)}$ and that $\overleftrightarrow{\tau_i}^{(3)} - \overleftrightarrow{\tau_i}^{(4)} = \overleftrightarrow{Q}_i^{(3)} - \overleftrightarrow{Q}_i^{(6)} + \frac{2}{3}(\overleftrightarrow{Q}_i^{(4)} - B^2\overleftrightarrow{Q}_i^{(1)} - \overleftrightarrow{Q}_i^{(5)})$. With these relations and Eqs. (8.5a-8.5f), it is readily seen that

$$
\begin{gathered}
(\vec{c}_i \times \vec{B}) \cdot \frac{\partial}{\partial \vec{c}_i} \overleftrightarrow{Q}_i^{(0)} = 0 \\
(\vec{c}_i \times \vec{B}) \cdot \frac{\partial}{\partial \vec{c}_i} \overleftrightarrow{Q}_i^{(1)} = 2\overleftrightarrow{Q}_i^{(2)} \\
(\vec{c}_i \times \vec{B}) \cdot \frac{\partial}{\partial \vec{c}_i} \overleftrightarrow{Q}_i^{(2)} = -B^2\overleftrightarrow{Q}_i^{(1)} + \overleftrightarrow{Q}_i^{(3)} + \overleftrightarrow{Q}_i^{(6)} \\
(\vec{c}_i \times \vec{B}) \cdot \frac{\partial}{\partial \vec{c}_i} \overleftrightarrow{Q}_i^{(3)} = -\frac{4}{3}B^2\overleftrightarrow{Q}_i^{(2)} + 2\overleftrightarrow{Q}_i^{(7)} \\
(\vec{c}_i \times \vec{B}) \cdot \frac{\partial}{\partial \vec{c}_i} \overleftrightarrow{Q}_i^{(4)} = (\vec{c}_i \times \vec{B}) \cdot \frac{\partial}{\partial \vec{c}_i} \overleftrightarrow{Q}_i^{(5)} = (\vec{c}_i \times \vec{B}) \cdot \frac{\partial}{\partial \vec{c}_i} \overleftrightarrow{Q}_i^{(8)} = 0 \\
(\vec{c}_i \times \vec{B}) \cdot \frac{\partial}{\partial \vec{c}_i} \overleftrightarrow{Q}_i^{(6)} = \overleftrightarrow{Q}_i^{(7)} - \frac{2}{3}B^2\overleftrightarrow{Q}_i^{(2)} \\
(\vec{c}_i \times \vec{B}) \cdot \frac{\partial}{\partial \vec{c}_i} \overleftrightarrow{Q}_i^{(7)} = \overleftrightarrow{Q}_i^{(8)} - B^2\overleftrightarrow{Q}_i^{(6)} + \frac{1}{3}(B^2\overleftrightarrow{Q}_i^{(4)} - B^4\overleftrightarrow{Q}_i^{(1)})
\end{gathered}
\tag{8.8}
$$

where use is made the fact that $\left(\frac{\partial}{\partial \vec{c}_i} Q_i^{(n)}\right) \cdot (\vec{c}_i \times \vec{B}) = 0$ for $n = 4, 5$ and 8. If instead of Eq. (8.3) we now propose that

$$
\overleftrightarrow{B} = \sum_{n=0}^{8} \Upsilon_i^n \overleftrightarrow{Q}_i^{(n)} \tag{8.9}
$$

then it is easy to show, equating Eqs. (8.3) and (8.9) and using Eqs. (8.7), that one may obtain the scalars Γ_i^n in terms of the Υ_i^n's. The result is that,

$$
\begin{gathered}
\Gamma_i^0 = \Upsilon_i^0 \\
\Upsilon_i^1 = \Gamma_i^1 - \frac{1}{3}B^2\Gamma_i^3 + \frac{1}{3}B^4\Gamma_i^4 \\
\Gamma_i^2 = \Upsilon_i^2 \\
\Gamma_i^3 = \Upsilon_i^3 \\
\Upsilon_i^4 = \frac{1}{3}(\Gamma_i^3 - \Gamma_i^4) \\
\Upsilon_i^5 = \frac{1}{3}(\Gamma_i^4 - \Gamma_i^3 + B^2\Gamma_i^6)
\end{gathered}
\tag{8.10}
$$

$$\Upsilon_i^6 = \Gamma_i^4$$
$$\Upsilon_i^7 = \Gamma_i^5$$
$$\Upsilon_i^8 = \Gamma_i^6$$

Furthermore, notice that when $(\vec{c}_i \times \vec{B}) \cdot \frac{\partial}{\partial \vec{c}_i}$ acts upon $\overleftrightarrow{\tau}_i^{(n)}$ it generates the same tensor multiplied by c_i^2 and/or B^2 never by $(\vec{c}_i \cdot \vec{B})^2$. Thus, the scalar functions that appear in Eqs. (8.3) or in Eq. (8.9) will not depend on this last term. This means that we may safely assume that the Υ_i^n's will not depend on this scalar.

We are now in a position to deal with the integral given by Eq. (8.4) since all tensors $\overleftrightarrow{Q}_i^n$ now contain the symmetric traceless tensor $\overleftrightarrow{c_i\,{}^0c_i}$. Substitution of Eq. (8.9) into Eq. (8.4) and collecting terms we get a set of nine integral equations for the Υ_i^n functions namely, $C(\Upsilon_i^0, \overleftrightarrow{c_i\,{}^0c_i}) = 0$ which implies $\Upsilon_i^0 = 0$ and,

$$
\begin{aligned}
\tfrac{m_i}{kT}\overleftrightarrow{c_i\,{}^0c_i} &= \tfrac{e_i}{m_i}\left(B^2\Upsilon_i^2 + \tfrac{1}{3}B^4\Upsilon_i^7\right)\overleftrightarrow{c_i\,{}^0c_i} + C(\Upsilon_i^1\overleftrightarrow{c_i\,{}^0c_i}) \\
0 &= \tfrac{e_i}{m_i}\left(\tfrac{2}{3}B^2\Upsilon_i^6 + \tfrac{4}{3}B^2\Upsilon_i^3 - 2\Upsilon_i^1\right)\overleftrightarrow{c_i\,{}^0c_i} + C(\Upsilon_i^2\overleftrightarrow{c_i\,{}^0c_i}) \\
0 &= -\tfrac{e_i}{m_i}\Upsilon_i^2\overleftrightarrow{c_i\,{}^0c_i} + C(\Upsilon_i^3\overleftrightarrow{c_i\,{}^0c_i}) \\
0 &= -\tfrac{1}{3}\tfrac{e_i}{m_i}B^2\Upsilon_i^7\overleftrightarrow{c_i\,{}^0c_i} + C(\Upsilon_i^4\overleftrightarrow{c_i\,{}^0c_i}) \\
0 &= C(\Upsilon_i^5, c_i^0c_i) \quad \Rightarrow \quad \Upsilon_i^5 = 0 \\
0 &= \tfrac{e_i}{m_i}\left(B^2\Upsilon_i^7 - \Upsilon_i^2\right)\overleftrightarrow{c_i\,{}^0c_i} + C(\Upsilon_i^6\overleftrightarrow{c_i\,{}^0c_i}) \\
0 &= -\tfrac{e_i}{m_i}\left(2\Upsilon_i^3 + \Upsilon_i^6\right)\overleftrightarrow{c_i\,{}^0c_i} + C(\Upsilon_i^7\overleftrightarrow{c_i\,{}^0c_i}) \\
0 &= -\tfrac{e_i}{m_i}\Upsilon_i^7\overleftrightarrow{c_i\,{}^0c_i} + C(\Upsilon_i^{(8)}\overleftrightarrow{c_i\,{}^0c_i})
\end{aligned}
\tag{8.11}
$$

Eqs. (8.11) is a set of eight linear coupled differential equations for the Υ_i^n functions ($\Upsilon_i^0 = 0$) where $C(\Upsilon_i^n\overleftrightarrow{c_i\,{}^0c_i})$ is an abbreviation for the full linearized collision term containing all interactions between the two species $i = a, b$ (see

Eq. (8.1)). Besides knowing already that also $\Upsilon_i^5 = 0$ as pointed out above, we now proceed to find out if the subsidiary conditions, namely

$$\sum_{i=a}^{b} m_i \int f_i^{(0)} \varphi_i^{(1)} \left\{ \begin{array}{c} 1 \\ \vec{c}_i \\ \frac{1}{2} c_i^2 \end{array} \right\} d\vec{c}_i = 0$$

impose further conditions on the Υ_i^n's. Since for our purposes

$$\varphi_i^{(1)} = \overleftrightarrow{B_i} : \text{grad } \vec{u}$$

and $\overleftrightarrow{B_i}$ is given by Eq. (8.9) we get that

$$\sum_{i=a}^{b} m_i \int f_i^{(0)} \sum_{n=0}^{8} \Upsilon_i^{(n)} \left\{ \begin{array}{c} 1 \\ \vec{c}_i \\ \frac{1}{2} c_i^2 \end{array} \right\} \overleftrightarrow{Q}_i^n d\vec{c}_i = 0$$

Now, $\Upsilon_i^{(0)} = 0$ and for the remaining terms all $\overleftrightarrow{Q}_i^n$ are proportional to $\overleftrightarrow{c_i \, {}^0 c_i}$ except $\overleftrightarrow{Q}_i^5 = \overleftrightarrow{B \, {}^0 B} c_i^2$. So except for this tensor all other integrals are odd functions of $\vec{c}_i$ and those who are not vanish since $\overleftrightarrow{c_i \, {}^0 c_i}$ is a traceless tensor that it reduces to

$$\sum_{i=a}^{b} m_i \int f_i^{(0)} \Upsilon_i^{(5)} \left\{ \begin{array}{c} 1 \\ \vec{c}_i \\ \frac{1}{2} c_i^2 \end{array} \right\} \overleftrightarrow{B \, {}^0 B} c_i^2 d\vec{c}_i = 0$$

This implies that to satisfy the vanishing of the integral even in $\vec{c}_i$, $\Upsilon_i^{(5)} = 0$ as already seen, which in turn implies that the set in Eq. (8.11) is reduced indeed to seven integral equations. Using this result and inverting Eqs. (8.10) we find that

$$\begin{gathered}
\Upsilon_i^2 + \tfrac{1}{3} B^2 \Upsilon_i^7 = \Gamma_i^{(2)} + \tfrac{1}{3} \Gamma_i^{(5)} \\
\tfrac{2}{3} B^2 \Upsilon_i^6 + \tfrac{4}{3} B^2 \Upsilon_i^3 - 2\Upsilon_i^1 = 2(B^2 \Gamma_i^{(3)} - \Gamma_i^{(1)}) \\
B^2 \Upsilon_i^{(7)} - \Upsilon_i^2 = B^2 \Gamma_i^{(5)} - \Gamma_i^{(2)} \\
2\Upsilon_i^3 - \Upsilon_i^4 = 2\Gamma_i^{(3)} + \Gamma_i^{(6)}
\end{gathered} \tag{8.12}$$

Eqs. (8.12) simply imply that solving for the Υ functions will allow the computation of the early Γ_i functions introduced in Eq. (8.3) which, as we shall see, are more convenient when determining the coupling of the stress tensor with the magnetic field. Writing the integral equations in terms of the Γ_i^n functions we easily find that,

$$\frac{m_i}{kT}\overleftrightarrow{c_i\,{}^0c_i} = \frac{e_i}{m_i}B^2(\Gamma_i^{(2)} + \frac{1}{3}B^2\Gamma_i^{(5)}) + C\left[\Gamma_i^{(1)} + \frac{1}{3}B^2(\Gamma_i^{(4)} - \Gamma_i^{(3)})\overleftrightarrow{c_i\,{}^0c_i}\right] \tag{8.13a}$$

$$0 = \frac{2e_i}{m_i}(B^2\Gamma_i^{(3)} - \Gamma_i^{(1)})\overleftrightarrow{c_i\,{}^0c_i} + C(\Gamma_i^{(2)}\overleftrightarrow{c_i\,{}^0c_i}) \tag{8.13b}$$

$$0 = \frac{e_i}{m_i}\Gamma_i^{(2)}\overleftrightarrow{c_i\,{}^0c_i} + C(\Gamma_i^{(3)}\overleftrightarrow{c_i\,{}^0c_i}) \tag{8.13c}$$

$$0 = \frac{-e_i}{3m_i}(B^2\Gamma_i^{(5)})\overleftrightarrow{c_i\,{}^0c_i} + C(\frac{1}{3}(\Gamma_i^{(3)} - \Gamma_i^{(4)})\overleftrightarrow{c_i\,{}^0c_i}) \tag{8.13d}$$

$$0 = \frac{e_i}{m_i}(B^2\Gamma_i^{(5)} - \Gamma_i^{(2)})\overleftrightarrow{c_i\,{}^0c_i} + C(\Gamma_i^{(4)}\overleftrightarrow{c_i\,{}^0c_i}) \tag{8.13e}$$

$$0 = \frac{-e_i}{m_i}(2\Gamma_i^{(3)} + \Gamma_i^{(4)})\overleftrightarrow{c_i\,{}^0c_i} + C(\Gamma_i^{(5)}\overleftrightarrow{c_i\,{}^0c_i}) \tag{8.13f}$$

$$0 = \frac{-e_i}{m_i}\Gamma_i^{(5)}\overleftrightarrow{c_i\,{}^0c_i} + C(\Gamma_i^{(6)}\overleftrightarrow{c_i\,{}^0c_i}) \tag{8.13g}$$

Moreover, using some straightforward transformations we may still reduce this set to three linear coupled integral equations. Indeed, multiplying Eq. (8.13g) by $\frac{1}{3}B^2$, remembering that C is a linear operator and comparing the resulting equation with Eq. (8.13d) we immediately see that

$$\Gamma_i^{(6)} = \frac{1}{B^2}(\Gamma_i^{(3)} - \Gamma_i^{(4)}) \tag{8.14a}$$

Using this result once more in Eq. (8.13g), yields

$$\frac{e_i}{3m_i}B^4\Gamma_i^{(5)}\overleftrightarrow{c_i\,{}^0c_i} = C\left[\frac{B^2}{3}(\Gamma_i^{(3)} - \Gamma_i^{(4)})\overleftrightarrow{c_i\,{}^0c_i}\right]$$

which when used in (8.13a) reduces this equation to a simpler one,

$$\frac{m_i}{kT}\overleftrightarrow{c_i\,{}^0c_i} = \frac{e_i}{m_i}B^2\Gamma_i^{(2)}\overleftrightarrow{c_i\,{}^0c_i} + C(\Gamma_i^{(1)}\overleftrightarrow{c_i\,{}^0c_i}) \tag{8.14b}$$

Substitution of Eq. (8.13e) into (8.13d) yields that

$$0 = -\frac{e_i}{m_i}\Gamma_i^{(2)}\overleftrightarrow{c_i{}^0c_i} + C(\Gamma_i^{(3)}\overleftrightarrow{c_i{}^0c_i}) \tag{8.14c}$$

Multiplying this equation by B^2 and add to Eq. (8.14b) to get

$$\frac{m_i}{kT}\overleftrightarrow{c_i{}^0c_i} = C(L_i\overleftrightarrow{c_i{}^0c_i}) \tag{8.15a}$$

where

$$L_i = \Gamma_i^{(1)} + B^2\Gamma_i^{(3)} \tag{8.15b}$$

Eqs. (8.15a,8.15b) are the first of the three sought equations. Notice now its immense simplicity. Now look at Eqs. (8.14b), (8.13b) and (8.14c). We rewrite (8.14b) as,

$$\frac{m_i}{kT}\overleftrightarrow{c_i{}^0c_i} = \frac{-iBe_i}{m_i}(i\Gamma_i^{(2)}B\overleftrightarrow{c_i{}^0c_i}) + C(\Gamma_i^{(1)}\overleftrightarrow{c_i{}^0c_i})$$

Now multiply Eq. (8.14c) by $-B^2 = (iB)(iB)$ and add to this one so that,

$$\frac{m_i}{kT}\overleftrightarrow{c_i{}^0c_i} = \frac{-2iBe_i}{m_i}(i\Gamma_i^{(2)}B\overleftrightarrow{c_i{}^0c_i}) + C\left[(i\Gamma_i^{(2)}B - B^2\Gamma_i^{(3)})\overleftrightarrow{c_i{}^0c_i}\right]$$

Multiplying Eq. (8.13b) by iB and adding the result to this one, yields,

$$\frac{m_i}{kT}\overleftrightarrow{c_i{}^0c_i} = \frac{-2iBe_i}{m_i}(\Gamma_i^{(1)} + i\Gamma_i^{(2)}B - B^2\Gamma_i^{(3)})\overleftrightarrow{c_i{}^0c_i} +$$

$$C\left[(\Gamma_i^{(1)} + i\Gamma_i^{(2)}B - B^2\Gamma_i^{(3)})\overleftrightarrow{c_i{}^0c_i}\right]$$

Define:

$$G_i = \Gamma_i^{(1)} + i\Gamma_i^{(2)}B - B^2\Gamma_i^{(3)} \tag{8.16a}$$

to get,

$$\frac{m_i}{kT}\overleftrightarrow{c_i{}^0c_i} = \frac{-2iBe_i}{m_i}G_i\overleftrightarrow{c_i{}^0c_i} + C(G_i\overleftrightarrow{c_i{}^0c_i}) \tag{8.16b}$$

Eq. (8.16b) is the second desired integral equation. Notice that from its solution,

$$\begin{aligned} Re\ G_i &= \Gamma_i^{(1)} - B^2\Gamma_i^{(3)} \\ Im\ G_i &= B\Gamma_i^{(2)} \end{aligned} \tag{8.17}$$

From Eqs. (8.15b) and (8.17),

$$\Gamma_i^{(1)} = (L_i)_{B=0} = (Re\ G_i)_B = 0$$

represents a consistency condition for the solutions. To obtain the last equation we rewrite Eq. (8.13f), after multiplication by B^2, as follows,

$$\frac{-2e_iB^2}{m_i}\Gamma_i^{(3)}\overleftrightarrow{c_i\,{}^0c_i} = \frac{e_iB^2}{m_i}\Gamma_i^{(4)} - C(B^2\Gamma_i^{(5)}\overleftrightarrow{c_i\,{}^0c_i})$$

If we now substitute this equation into Eq. (8.13b) multiplied by iB, we get that

$$0 = \frac{-iBe_i}{m_i}(2\Gamma_i^{(1)} + B^2\Gamma_i^{(4)})\overleftrightarrow{c_i\,{}^0c_i} + C\left[(iB^2\Gamma_i^{(2)} + B^2\Gamma_i^{(5)})\overleftrightarrow{c_i\,{}^0c_i}\right]$$

Multiply Eq. (8.13e) by B^2 and add the result to this equation to get

$$0 = \frac{-iBe_i}{m_i}(2\Gamma_i^{(1)} - iB\Gamma_i^{(2)} + B^2\Gamma_i^{(4)} + iB^3\Gamma_i^{(5)})\overleftrightarrow{c_i\,{}^0c_i}+$$

$$C\left[(B^2\Gamma_i^{(4)} + iB\Gamma_i^{(2)} + iB^3\Gamma_i^{(5)})\overleftrightarrow{c_i\,{}^0c_i}\right]$$

Now we multiply Eq. (8.14b) by Eq. (8.2) and add it to this equation to get finally that

$$\frac{2m_i}{kT}\overleftrightarrow{c_i\,{}^0c_i} = \frac{-iBe_i}{m_i}P_i\overleftrightarrow{c_i\,{}^0c_i} + C(P_i\overleftrightarrow{c_i\,{}^0c_i}) \tag{8.18a}$$

where

$$P_i = 2\Gamma_i^{(1)} + B^2\Gamma_i^{(4)} + iB(\Gamma_i^{(2)} + B^2\Gamma_i^{(5)}) \tag{8.18b}$$

Eqs. (8.18a) and (8.18b) are the third desired equations. Notice that,

$$Re\ P_i = 2\Gamma_i^{(1)} + B^2\Gamma_i^{(4)}$$

$$Im\ P_i = B(\Gamma_i^{(2)} + B^2\Gamma_i^{(5)}) \tag{8.19}$$

$$\Gamma_i^{(6)} = \tfrac{1}{B^2}(\Gamma_i^{(3)} - \Gamma_i^{(4)})$$

which together with Eqs. (8.15b) and (8.17) determine completely the six unknown functions $\Gamma_i^{(n)}$ (since $\Gamma_i^{(0)} = 0$). Furthermore, examination of the integral equations (8.15a), (8.16b) and (8.18a) clearly indicates that we only need to solve the generic equation (8.16b). Indeed, Eq. (8.18a) reads,

$$\frac{m_i}{kT}\overleftrightarrow{c_i{}^0c_i} = \frac{-iBe_i}{2m_i}P_i\overleftrightarrow{c_i{}^0c_i} + C(\frac{1}{2}P_i\overleftrightarrow{c_i{}^0c_i}) \tag{8.20}$$

which is identical to (8.16b) if in equation (8.20) $\frac{1}{2}P_i$ is identified with G_i and B is changed to $2B$. On the other hand Eq. (8.15a) follows from (8.20) setting $B = 0$ and $L_i \equiv G_i = \frac{1}{2}P_i$. Thus we have reduced the problem to the solution of a single linear integral equation, equation (8.21) which is readily solved by the same variational method described in Chapter 4.

Nevertheless before embarking in this calculation it is very convenient to first study the general structure of the stress tensor. As we shall see, this will simplify the task of solving Eq. (8.21),

$$\frac{m_i}{kT}\overleftrightarrow{c_i{}^0c_i} = \frac{-2iBe_i}{m_i}G_i\overleftrightarrow{c_i{}^0c_i} + C(G_i\overleftrightarrow{c_i{}^0c_i}) \tag{8.21}$$

8.2 The Stress Tensor

As we pointed out in Eq. (2.18), the stress tensor for the plasma is defined as

$$\overleftrightarrow{\tau} = \sum_{i=a}^{b} m_i \int f_i\vec{c}_i\vec{c}_i d\vec{c}_i$$

Substitution of Eq. (3.13) into this definition, noticing that $\vec{\mathbb{A}}_i$ and $\vec{\mathbb{D}}_i$ are all linear functions of $\vec{c}_i$ times even scalar functions of this variable, consistently with Curie's principle the terms in grad T and $\vec{d}_{ij}$ vanish and we are thus left with

$$\overleftrightarrow{\tau}' = \sum_{i=a}^{b} m_i \int f_i^{(0)}\vec{c}_i\vec{c}_i d\vec{c}_i \overleftrightarrow{\mathbb{B}}_i : \text{grad } \vec{u} \tag{8.22}$$

Using now the proposed expansion for $\overleftrightarrow{\mathbb{B}}_i$ given in Eq. (8.3) we obtain that

$$\overleftrightarrow{\tau}' = \sum_{i=a}^{b} m_i \int d\vec{c}_i f_i^{(0)}\vec{c}_i\vec{c}_i \sum_{n=1}^{6} \Gamma_i^n \overleftrightarrow{\tau}_i^n : \text{grad } \vec{u} \tag{8.23}$$

where $\overleftrightarrow{\tau}' = \overleftrightarrow{\tau} - p\mathbb{I}$, p being the hydrostatic pressure. Eq. (8.23) indicates that the evaluation of each term in the integral has to be performed individually for each of the tensors $\overleftrightarrow{\tau}_i^n$ listed in Eq. (8.2). Now, the scalar functions

Γ_i^n which are unknown, are now expanded in a complete set of orthonormal functions the Sonine polynomials of order 5/2. Thus,

$$\Gamma_i^n = \sum_{p=0}^{\infty} \alpha_i^{(p)(n)} S_{\frac{5}{2}}^{(p)}(c_i^2) \tag{8.24}$$

where the $\alpha_i^{(p)(n)}$ will be functions of the scalars n, T, B, c_i^2 and so on. Substitution of Eq. (8.24) in Eq. (8.23) apparently complicates things, since

$$\overleftrightarrow{\tau}' = \sum_{i=a}^{b} m_i \sum_{n=1}^{6} \int d\vec{c}_i f_i^{(0)} \vec{c}_i\vec{c}_i \sum_{p=0}^{\infty} \alpha_i^{(p)(n)} S_{\frac{5}{2}}^{(p)}(c_i^2) \overleftrightarrow{\tau}_i^n : \text{grad } \vec{u} \tag{8.25}$$

From the structure of the tensors $\overleftrightarrow{\tau}_i^n$ we see that all of them are bilinear functions of $\vec{c}_i$ whence they have the form of $|\vec{c}_i|$ times some angular dependence. The same happens with the dyad $\overleftrightarrow{c_i c_i} \equiv c_i^2 \times$ angular dependence. Thus the integral over c (magnitude) is the same for all terms in Eq. (8.25) and has the form

$$\int_0^\infty c^6 f_i^{(0)} S_{\frac{5}{2}}^{(p)} d\vec{c}_i = n_i \left(\frac{m_i}{2\pi kT}\right)^{\frac{3}{2}} \left(\frac{2kT}{m_i}\right)^{\frac{7}{2}} \int_0^\infty w_i^6 e^{-w_i^2} S_{\frac{5}{2}}^{(p)} S_{\frac{5}{2}}^{(0)} dw_i$$

when using dimensionless velocities (c.f. Eq. (3.29)). But by the properties of the Sonine polynomials we finally get that

$$\int_0^\infty c^6 f_i^{(0)} S_{\frac{5}{2}}^{(p)} d\vec{c}_i = n_i \left(\frac{2kT}{m_i}\right)^2 \frac{15}{16\pi}\delta_{p,0}$$

so that Eq. (8.25) now reads,

$$\overleftrightarrow{\tau}' = \sum_{i=a}^{b} \frac{n_i}{m_i}(2kT)^2 \frac{15}{16\pi} \sum_{n=1}^{6} \alpha_i^{(0)(n)} \int\int d\Omega (\overleftrightarrow{c_i c_i})_\Omega (\overleftrightarrow{\tau}_i^n)_\Omega : \text{grad } \vec{u} \tag{8.26}$$

where $d\Omega = \sin\theta d\theta d\varphi$ and the subscript Ω means taking the angular part only of the involved quantities. Moreover the infinite number of terms involved in Eq. (8.24) for the $\Gamma_i^{(n)}$'s reduce to only six coefficients for each species since $\Gamma_i^{(n)} = \alpha_i^{(0)(n)}$. This considerably simplifies all calculations. Now the rest of the calculations are a simple matter of algebra.

Let us define,

$$\overleftrightarrow{I}^{n} = \frac{15}{4\pi}\alpha_i^{(0)(n)} \int_0^{2\pi} d\varphi \int_0^{\pi} \sin\theta d\theta (\vec{c}_i\vec{c}_i)_\Omega (\overleftrightarrow{\tau}_i^{n'})_\Omega : \text{grad } \vec{u}$$

In general

$$\begin{aligned}\overleftrightarrow{\tau}_i^{n} : \text{grad } \vec{u} &= (\overleftrightarrow{\tau}_i^{n})_{xx}\frac{\partial u_x}{\partial x} + (\overleftrightarrow{\tau}_i^{n})_{xy}\frac{\partial u_x}{\partial y} + (\overleftrightarrow{\tau}_i^{n})_{xz}\frac{\partial u_x}{\partial z} \\ &+(\overleftrightarrow{\tau}_i^{n})_{xy}\frac{\partial u_y}{\partial x} + (\overleftrightarrow{\tau}_i^{n})_{yy}\frac{\partial u_y}{\partial y} + (\overleftrightarrow{\tau}_i^{n})_{zy}\frac{\partial u_y}{\partial z} \\ &+(\overleftrightarrow{\tau}_i^{n})_{xz}\frac{\partial u_z}{\partial x} + (\overleftrightarrow{\tau}_i^{n})_{yz}\frac{\partial u_z}{\partial y} + (\overleftrightarrow{\tau}_i^{n})_{zz}\frac{\partial u_z}{\partial z}\end{aligned}$$

and $(\overleftrightarrow{\tau}_i^{n})_{kl} = (\overleftrightarrow{\tau}_i^{n})_{lk}$ since it is a symmetric tensor.

Taking $n = 1$, $\overleftrightarrow{\tau}_i^{1} = \overleftrightarrow{c_i{}^0c_i} = \vec{c}_i\vec{c}_i + \frac{1}{3}\mathbb{I}$

$$\begin{aligned}\overleftrightarrow{\tau}_i^{n} : \text{grad } \vec{u} &= \left(c_x^2 - \tfrac{1}{3}c^2\right)\tfrac{\partial u_x}{\partial x} + c_xc_y\left(\tfrac{\partial u_x}{\partial y} + \tfrac{\partial u_y}{\partial x}\right) + c_xc_z\left(\tfrac{\partial u_x}{\partial z} + \tfrac{\partial u_z}{\partial x}\right) \\ &+ \left(c_y^2 - \tfrac{1}{3}c^2\right)\tfrac{\partial u_y}{\partial y} + c_yc_z\left(\tfrac{\partial u_y}{\partial z} + \tfrac{\partial u_z}{\partial y}\right) + \left(c_z^2 - \tfrac{1}{3}c^2\right)\tfrac{\partial u_z}{\partial z}\end{aligned}$$

Now,

$$(\overleftrightarrow{I}^{(1)})_{xx} = \tfrac{15}{4\pi}\alpha_i^{(0)(n)} \int_0^{2\pi} d\varphi \int_0^{\pi} \sin\theta d\theta \cos^2\varphi \sin^2\theta$$

$$\times\left\{ \left(\cos^2\varphi\sin^2 - \tfrac{1}{3}\right)\tfrac{\partial u_x}{\partial x} + \left(\sin^2\varphi\sin\theta^2 - \tfrac{1}{3}\right)\tfrac{\partial u_y}{\partial y} + \left(\cos^2\theta - \tfrac{1}{3}\right)\tfrac{\partial u_z}{\partial z}\right\}$$

using spherical coordinates. Using further the auxiliary table of integrals in Appendix G we see that,

$$(\overleftrightarrow{I}^{(1)})_{xx} = \alpha_i^{(0)(1)}2\left(-\frac{1}{3}\text{div } \vec{u} + \frac{\partial u_x}{\partial x}\right)$$

By an identical procedure,

$$(\overleftrightarrow{I}^{(1)})_{yy} = \alpha_i^{(0)(1)}2\left(-\frac{1}{3}\text{div } \vec{u} + \frac{\partial u_y}{\partial y}\right)$$

$$(\overleftrightarrow{I}^{(1)})_{zz} = \alpha_i^{(0)(1)}2\left(-\frac{1}{3}\text{div } \vec{u} + \frac{\partial u_z}{\partial z}\right)$$

Now,

$$\begin{aligned}(\overleftrightarrow{I}^{1})_{xy} &= \tfrac{15}{4\pi}\alpha_i^{(0)(1)} \int_0^{2\pi} d\varphi \int_0^{\pi} d\theta \sin\theta \cos^2\varphi \sin^2\varphi \sin^2\theta \left(\tfrac{\partial u_x}{\partial y} + \tfrac{\partial u_y}{\partial x}\right) \\ &= \alpha_i^{(0)(1)}\left(\tfrac{\partial u_x}{\partial y} + \tfrac{\partial u_y}{\partial x}\right)\end{aligned}$$

Also,

$$(\overleftrightarrow{I}^1)_{xz} = \tfrac{15}{4\pi}\alpha_i^{(0)(1)} \int_0^{2\pi} d\varphi \int_0^{\pi} d\theta \sin\theta \cos^2\varphi \sin^2\theta \cos^2\theta \left(\tfrac{\partial u_x}{\partial z} + \tfrac{\partial u_z}{\partial x}\right)$$
$$= \alpha_i^{(0)(1)} \left(\tfrac{\partial u_x}{\partial z} + \tfrac{\partial u_z}{\partial x}\right)$$

and similarly for the yz component. Thus, $\overleftrightarrow{\tau} - p\mathbb{I}$ for $n = 1$ is determined by these six quantities.

From the structure of $\overleftrightarrow{\tau}^{(2)}$ in Eq. (8.2) we see that

$$\tfrac{1}{B}\overleftrightarrow{\tau}^{(2)} : \text{grad}\,\vec{u} = c_{ix}c_{iy}\tfrac{\partial u_x}{\partial x} - \tfrac{1}{2}\left(c_{ix}^2 - c_{iy}^2\right)\left(\tfrac{\partial u_x}{\partial y} + \tfrac{\partial u_y}{\partial x}\right)$$
$$+\tfrac{1}{2}c_{iz}c_{iy}\left(\tfrac{\partial u_x}{\partial z} + \tfrac{\partial u_z}{\partial x}\right) + c_{iy}c_{ix}\tfrac{\partial u_y}{\partial y} - \tfrac{1}{2}c_{iz}c_{ix}\left(\tfrac{\partial u_y}{\partial z} + \tfrac{\partial u_z}{\partial y}\right)$$

The only non vanishing terms in $(\overleftrightarrow{I}^{(2)})$ are $(\overleftrightarrow{I}^{(2)})_{xx}$, $(\overleftrightarrow{I}^{(2)})_{yy}$, $(\overleftrightarrow{I}^{(2)})_{zz}$, $(\overleftrightarrow{I}^{(2)})_{xy}$, $(\overleftrightarrow{I}^{(2)})_{zy}$ and $(\overleftrightarrow{I}^{(2)})_{zx}$.

$$(\overleftrightarrow{I}^2)_{xx} = \tfrac{15}{4\pi}\alpha_i^{(0)(2)} B \int_0^{2\pi} d\varphi \int_0^{\pi} d\theta \sin\theta \tfrac{1}{2}\cos^2\varphi \sin^2\theta$$
$$(\cos^2\varphi\sin^2\theta - \sin^2\varphi\cos^2\theta)\left(\tfrac{\partial u_x}{\partial y} + \tfrac{\partial u_y}{\partial x}\right)$$
$$= -\alpha_i^{(0)(2)} B\left(\tfrac{\partial u_x}{\partial y} + \tfrac{\partial u_y}{\partial x}\right)$$

$$(\overleftrightarrow{I}^2)_{yy} = \tfrac{15}{4\pi}\alpha_i^{(0)(2)} B \int_0^{2\pi} d\varphi \int_0^{\pi} d\theta \sin\theta (\tfrac{1}{2}\sin^2\varphi \sin^2\theta)$$
$$(\cos^2\varphi\sin^2\theta - \sin^2\varphi\cos^2\theta)\left(\tfrac{\partial u_x}{\partial y} + \tfrac{\partial u_y}{\partial x}\right)$$
$$= \alpha_i^{(0)(2)} B\left(\tfrac{\partial u_y}{\partial x} + \tfrac{\partial u_x}{\partial y}\right)$$

$$(\overleftrightarrow{I}^2)_{zz} = \tfrac{15}{4\pi}\alpha_i^{(0)(2)} B \int_0^{2\pi} d\varphi \int_0^{\pi} d\theta \sin\theta$$
$$(-\tfrac{1}{2})(\cos^2\varphi\sin^2\theta - \sin^2\varphi\cos^2\theta)\left(\tfrac{\partial u_x}{\partial y} + \tfrac{\partial u_y}{\partial x}\right)$$
$$= -\tfrac{15}{8\pi}\tfrac{4}{15}\alpha_i^{(0)(2)} B \int_0^{2\pi} d\varphi(\cos^2\varphi - \sin^2\varphi) = 0$$

$$(\overleftrightarrow{I}^2)_{xy} = \tfrac{15}{8\pi}\alpha_i^{(0)(2)} B \int_0^{2\pi} d\varphi \int_0^{\pi} d\theta \sin\theta \cos^2\varphi \sin^2\theta \sin^2\varphi \sin^2\theta \left(\tfrac{\partial u_x}{\partial y} - \tfrac{\partial u_y}{\partial x}\right)$$
$$= \tfrac{1}{2}\alpha_i^{(0)(2)} B\left(\tfrac{\partial u_x}{\partial y} - \tfrac{\partial u_y}{\partial x}\right)$$

$$(\overleftrightarrow{I}^2)_{xz} = -\tfrac{15}{8\pi}\alpha_i^{(0)(2)} B \int_0^{2\pi} d\varphi \int_0^{\pi} d\theta \sin\theta \tfrac{1}{2}\cos^2\theta \sin^2\theta \cos^2\varphi$$
$$= -\tfrac{1}{2}\alpha_i^{(0)(2)}\left(\tfrac{\partial u_y}{\partial z} + \tfrac{\partial u_z}{\partial y}\right)$$

Take $n = 3$. From the structure of $\overleftrightarrow{\tau}^3$ we see immediately that

$$\begin{aligned}\overleftrightarrow{\tau}_i^{(3)} : \text{grad } \vec{u} = &\ \tfrac{1}{3}\left(2c_{iy}^2 - c_{ix}^2\right) B^2 \tfrac{\partial u_x}{\partial x} - c_{ix} c_{iy} B^2 \left(\tfrac{\partial u_x}{\partial y} + \tfrac{\partial u_y}{\partial x}\right) \\ &+ \tfrac{1}{3}\left(2c_{ix}^2 - c_{iy}^2\right) B^2 \tfrac{\partial u_y}{\partial y} - \tfrac{1}{3}\left(2c_{ix}^2 - c_{iy}^2\right) B^2 \tfrac{\partial u_z}{\partial z}\end{aligned}$$

Using the fact that $c_{ix}^2 + c_{iy}^2 = c^2 - c_{iz}^2$ and carrying out the integrals over the angles it is now straightforward to see that

$$\begin{aligned}(\overleftrightarrow{I}^3)_{xx} &= \alpha_i^{(0)(3)} B^2 \left(-\tfrac{1}{3}\text{div } \vec{u} + \tfrac{5}{3}\left(\tfrac{\partial u_y}{\partial y} - \tfrac{\partial u_z}{\partial z}\right)\right) \\ (\overleftrightarrow{I}^3)_{yy} &= \tfrac{1}{3}\alpha_i^{(3)(0)} B^2 \left(5\tfrac{\partial u_x}{\partial x} - \tfrac{\partial u_y}{\partial y} - \tfrac{\partial u_z}{\partial z}\right) \\ (\overleftrightarrow{I}^3)_{zz} &= 3\alpha_i^{(3)(0)} B^2 \left(\tfrac{1}{3}\text{div } \vec{u} - \tfrac{\partial u_z}{\partial z}\right) \\ (\overleftrightarrow{I}^3)_{xy} &= -\alpha_i^{(3)(0)} B^2 \left(\tfrac{\partial u_x}{\partial y} + \tfrac{\partial u_y}{\partial x}\right)\end{aligned}$$

all other components are zero.

For $n = 4$,

$$\begin{aligned}\tfrac{1}{B^2}\overleftrightarrow{\tau}_i^{(4)} : \text{grad } \vec{u} = &-\tfrac{1}{3}c_{iz}^2 \text{div } \vec{u} + c_{iz}^2 \tfrac{\partial u_z}{\partial z} + \tfrac{1}{2}c_{ix}c_{iz}\left(\tfrac{\partial u_x}{\partial z} + \tfrac{\partial u_z}{\partial x}\right) \\ &+ \tfrac{1}{2}c_{iy}^2 c_{iz}^2 \left(\tfrac{\partial u_y}{\partial z} + \tfrac{\partial u_z}{\partial y}\right)\end{aligned}$$

Once more after integrating over the angles we get that

$$\begin{aligned}(\overleftrightarrow{I}^4)_{xx} &= \alpha_i^{(0)(4)} B^2 \left(-\tfrac{1}{3}\text{div } \vec{u} + \tfrac{\partial u_z}{\partial z}\right) \\ (\overleftrightarrow{I}^4)_{xz} &= \tfrac{1}{2}\alpha_i^{(0)(4)} B^2 \left(\tfrac{\partial u_x}{\partial z} + \tfrac{\partial u_z}{\partial x}\right) \\ (\overleftrightarrow{I}^4)_{yz} &= \tfrac{1}{2}\alpha_i^{(0)(4)} B^2 \left(\tfrac{\partial u_y}{\partial z} + \tfrac{\partial u_z}{\partial y}\right) \\ (\overleftrightarrow{I}^4)_{yy} &= \alpha_i^{(0)(4)} B^2 \left(-\tfrac{1}{3}\text{div } \vec{u} + \tfrac{\partial u_z}{\partial z}\right) = \tfrac{1}{3}(\overleftrightarrow{I}^4)_{zz}\end{aligned}$$

For $n = 5$ we have that

$$\overleftrightarrow{\tau}^{(5)} : \text{grad } \vec{u} = \tfrac{1}{2}B^3 c_{iz} c_{iy}\left(\tfrac{\partial u_x}{\partial z} + \tfrac{\partial u_z}{\partial x}\right) - \tfrac{1}{2}B^2 c_{iz} c_{ix}\left(\tfrac{\partial u_y}{\partial z} + \tfrac{\partial u_z}{\partial y}\right)$$

so that

$$\begin{aligned}(\overleftrightarrow{I}^5)_{yz} &= \tfrac{1}{2}\alpha_i^{(0)(5)} B^3 \left(\tfrac{\partial u_x}{\partial z} + \tfrac{\partial u_z}{\partial x}\right) \\ (\overleftrightarrow{I}^5)_{xz} &= -\tfrac{1}{2}\alpha_i^{(0)(5)} B^3 \left(\tfrac{\partial u_y}{\partial z} + \tfrac{\partial u_z}{\partial y}\right)\end{aligned}$$

and all other terms are zero by parity considerations.

Finally, for $n = 6$ we have that,

$$\begin{aligned}\overleftrightarrow{\tau}^{(6)} : \text{grad}\, \vec{u} &= -\tfrac{1}{3} B^4 c_{iz}^2 \left(\tfrac{\partial u_x}{\partial x} + \tfrac{\partial u_y}{\partial y}\right) + \tfrac{2}{3} B^4 c_{iz}^2 \left(\tfrac{\partial u_z}{\partial z}\right) \\ &= B^4 c_{iz}^2 \left(-\tfrac{1}{3}\text{div}\, \vec{u} + \tfrac{\partial u_z}{\partial z}\right)\end{aligned}$$

so that,

$$\begin{aligned}(\overleftrightarrow{I}^{6})_{xx} &= \alpha_i^{(0)(6)} B^4 \left(-\tfrac{1}{3}\text{div}\, \vec{u} + \tfrac{\partial u_z}{\partial z}\right) \\ (\overleftrightarrow{I}^{6})_{yy} &= \alpha_i^{(0)(6)} B^4 \left(-\tfrac{1}{3}\text{div}\, \vec{u} + \tfrac{\partial u_z}{\partial z}\right) \\ (\overleftrightarrow{I}^{6})_{zz} &= 3\alpha_i^{(0)(6)} B^4 \left(-\tfrac{1}{3}\text{div}\, \vec{u} + \tfrac{\partial u_z}{\partial z}\right)\end{aligned}$$

Now we collect terms in Eq. (8.26),

$$(\overleftrightarrow{\tau} - p\mathbb{I}) = \sum_{i=a}^{b} \frac{n_i}{m_i} (kT)^2 \sum_{n=1}^{6} \left(\overleftrightarrow{I}^{(n)}\right)$$

and we proceed in each case by components. Thus from the results in all previous pages,

$$\begin{aligned}(\overleftrightarrow{\tau} - p\mathbb{I})_{xx} &= \textstyle\sum_{i=a}^{b} \frac{n_i}{m_i} (kT)^2 \Big\{ 2\alpha_i^{(1)} \overleftrightarrow{S}_{xx} - 2\alpha_i^{(2)} B \overleftrightarrow{S}_{xy} \\ &+ \alpha_i^{(3)} B^2 (2\overleftrightarrow{S}_{yy} - \overleftrightarrow{S}_{zz}) + \alpha_i^{(4)} B^2 \overleftrightarrow{S}_{zz} + \alpha_i^{(6)} B^4 \overleftrightarrow{S}_{zz} \Big\}\end{aligned} \tag{8.26a}$$

where we have defined the tensor $\overleftrightarrow{S}$ as follows,

$$\overleftrightarrow{S}_{\alpha\beta} = \frac{1}{2}\left(\frac{\partial u_\alpha}{\partial x_\beta} + \frac{\partial u_\beta}{\partial x_\alpha}\right) - \frac{1}{3}\text{div}\, \vec{u}\, \delta_{\alpha\beta} \tag{8.26b}$$

$$\begin{aligned}(\overleftrightarrow{\tau} - p\mathbb{I})_{yy} &= \textstyle\sum_{i=a}^{b} \frac{n_i}{m_i} (kT)^2 \Big\{ 2\alpha_i^{(1)} \overleftrightarrow{S}_{yy} + 2\alpha_i^{(2)} B \overleftrightarrow{S}_{xy} \\ &+ \alpha_i^{(3)} B^2 (2\overleftrightarrow{S}_{xx} - \overleftrightarrow{S}_{zz}) + \alpha_i^{(4)} B^2 \overleftrightarrow{S}_{zz} + \alpha_i^{(6)} B^4 \overleftrightarrow{S}_{zz} \Big\}\end{aligned} \tag{8.26c}$$

$$\begin{aligned}(\overleftrightarrow{\tau} - p\mathbb{I})_{zz} &= \textstyle\sum_{i=a}^{b} \frac{n_i}{m_i} (kT)^2 \Big\{ 2\alpha_i^{(1)} \overleftrightarrow{S}_{zz} - 3\alpha_i^{(3)} B^2 \overleftrightarrow{S}_{zz} \\ &+ 3\alpha_i^{(4)} B^2 \overleftrightarrow{S}_{zz} + 3\alpha_i^{(6)} B^4 \overleftrightarrow{S}_{zz} \Big\}\end{aligned} \tag{8.26d}$$

$$\overleftrightarrow{\tau}_{xy} = \sum_{i=a}^{b} \frac{n_i}{m_i}(kT)^2 \Big\{ 2\alpha_i^{(1)} \overleftrightarrow{S}_{xy} + \alpha_i^{(2)} B(\overleftrightarrow{S}_{xx} - \overleftrightarrow{S}_{yy}) + 2\alpha_i^{(3)} B^2 (2\overleftrightarrow{S}_{xy} \Big\} \tag{8.27a}$$

$$\begin{aligned}\overleftrightarrow{\tau}_{xz} = \sum_{i=a}^{b} \frac{n_i}{m_i}(kT)^2 \Big\{ & 2\alpha_i^{(1)} \overleftrightarrow{S}_{xz} - \alpha_i^{(2)} B \overleftrightarrow{S}_{xz} \\ & + \alpha_i^{(4)} B^2 \overleftrightarrow{S}_{xz} - \alpha_i^{(5)} B^3 \overleftrightarrow{S}_{yz} \Big\}\end{aligned} \tag{8.27b}$$

$$\begin{aligned}\overleftrightarrow{\tau}_{yz} = \sum_{i=a}^{b} \frac{n_i}{m_i}(kT)^2 \Big\{ & 2\alpha_i^{(1)} \overleftrightarrow{S}_{yz} + \alpha_i^{(4)} B^2 \overleftrightarrow{S}_{yz} \\ & + 2\alpha_i^{(4)} B \overleftrightarrow{S}_{xz} - \alpha_i^{(5)} B^3 \overleftrightarrow{S}_{xz} \Big\}\end{aligned} \tag{8.27c}$$

These are the six component of the stress tensor in the presence of a magnetic field. The superscript (0) in the α's has been omitted since it is no longer necessary and clearly, $\alpha_i^{(n)} = \Gamma_i^{(n)}$ in Eq. (8.24) since only the first term in the summation survives. So there are twelve coefficients α_i to be determined, six for each species which will be obtained by solving Eq. (2.21). Notice also that when $B = 0$ only $\alpha_i^{(1)}$ survives which corresponds to the shear viscosity of a simple dilute mixture. Indeed,

$$\eta = \sum_{i=a}^{b} 2(kT)^2 \frac{n_i}{m_i} \alpha_i^{(1)} \tag{8.28}$$

a well known result. There is no bulk viscosity for a dilute mixture, one needs the existence of collisional transfer. Also it is important to notice that since $-\frac{1}{B}\alpha_i^{(6)} = \alpha_i^{(4)} - \alpha_i^{(3)}$ the last three terms in the bracket in the zz component cancel out so that

$$(\overleftrightarrow{\tau} - p\mathbb{I})_{zz} = \sum_{i=a}^{b} \frac{n_i}{m_i}(kT)^2 2\alpha_i^{(1)} \overleftrightarrow{S}_{zz} \tag{8.26d}$$

the stress tensor is not modified along the direction of the field.

8.3 The Integral Equation

To solve Eq. (8.21) we shall proceed in exactly the same way as we did with Eq. (5.7) free from the restrictions imposed by the subsidiary conditions as we mentioned in Section 8.1. This allows for the direct constructions of Davison's function for which purpose we multiply by $\mathfrak{I}_i \overleftrightarrow{c_i{}^0 c_i} f_i^{(0)}$, sum over i and integrate over $\vec{c}_i$. This yields

$$\mathfrak{D}(\mathfrak{I}_i) = \sum_{i=a}^{b} \frac{m_i}{kT} \int d\vec{c}_i f_i^{(0)} \mathfrak{I}_i \Big\{ - \overleftrightarrow{c_i{}^0 c_i} : \overleftrightarrow{c_i{}^0 c_i} - 2i \frac{e_i}{m_i} B \mathfrak{I}_i \overleftrightarrow{c_i{}^0 c_i} : \overleftrightarrow{c_i{}^0 c_i} +$$

$$C(\mathfrak{I}_i \overleftrightarrow{c_i{}^0 c_i}) + C(\mathfrak{I}_i \overleftrightarrow{c_i{}^0 c_i} + \mathfrak{I}_j \overleftrightarrow{c_j{}^0 c_j}) \Big\} \tag{8.29}$$

We now propose that

$$\mathfrak{I}_i \equiv G_i = \sum_{q=o}^{M} g_i^{(q)} S_{\frac{5}{2}}^{(q)}(c_i^2) \tag{8.30}$$

and noticing that $\overleftrightarrow{c_i{}^0 c_i} : \overleftrightarrow{c_i{}^0 c_i} = \frac{2}{3} c_i^4$ as may be easily verified we may calculate the first two integrals namely,

$$- \int d\vec{c}_i f_i^{(0)} \frac{2}{3} c_i^4 \sum_{q=o}^{M} g_i^{(q)} S_{\frac{5}{2}}^{(q)} = -10 \frac{n_i}{m_i^2} (kT)^2 g_i^{(0)}$$

whereas

$$2i \frac{e_i}{m_i} B \int d\vec{c}_i f_i^{(0)} \frac{2}{3} c_i^4 \sum_{q=o}^{M} g_i^{(q)} S_{\frac{5}{2}}^{(p)} S_{\frac{5}{2}}^{(q)} = -20 \omega_i \frac{n_i}{m_i^2} (kT)^2 (g_i^{(0)})^2$$

since the integral is only different from zero if $p = q$. Here, $\omega_i = \frac{e_i}{m_i} B$ as usual. Substituting these results in Eq. (8.29) and writing the collision kernel in full, we get that

$$\tfrac{1}{2}\mathfrak{D}(\mathfrak{I}_i) = -5kT \left(\tfrac{n_a}{m_a} g_a^{(0)} + \tfrac{n_b}{m_b} g_b^{(0)} \right) - 10i(kT)^2 B \left(\tfrac{e_a n_a}{m_a^3} (g_a^{(0)})^2 + \tfrac{e_b n_b}{m_b^3} (g_b^{(0)})^2 \right)$$

$$\tfrac{1}{2} \textstyle\sum_{i,j} \int \cdots \int f_i^{(0)} f_j^{(0)} d\vec{v}_i' d\vec{c}_j \sigma(\Omega) d\Omega g_{ji} \{ g_i^{(0)} \overleftrightarrow{c_i{}^0 c_i} \}$$

$$\Big[\{ g_i^{(0)} \overleftrightarrow{c_i{}^0 c_i} \} + \{ g_i^{(0)} \overleftrightarrow{c_i{}^0 c_i} + g_j^{(0)} \overleftrightarrow{c_j{}^0 c_j} \} \Big] \tag{8.31}$$

where have retained only the first term in the series (8.30). Using exactly the same arguments of part I in dealing with the collision integrals, identifying the function G_{ij} of Eq. (5.4a), we find that to a first approximation the third term in (8.31) reads as

$$4(kT)^2\left\{\frac{n_a^2}{m_a^2}(g_a^{(0)})^2\left[\overleftrightarrow{w_a\,{}^0w_a},\overleftrightarrow{w_a\,{}^0w_a}\right]_{aa}+\frac{n_b^2}{m_b^2}(g_b^{(0)})^2\left[\overleftrightarrow{w_b\,{}^0w_b},\overleftrightarrow{w_b\,{}^0w_b}\right]_{bb}+\right.$$

$$\frac{n_an_b}{2m_am_b}\left\{\left[\overleftrightarrow{w_a\,{}^0w_a},\overleftrightarrow{w_a\,{}^0w_a}\right]_{ab}(g_a^{(0)})^2+\left[\overleftrightarrow{w_b\,{}^0w_b},\overleftrightarrow{w_b\,{}^0w_b}\right]_{ab}(g_b^{(0)})^2+\right.$$

$$\left.\left.2g_a^{(0)}g_b^{(0)}\left[\overleftrightarrow{w_a\,{}^0w_a},\overleftrightarrow{w_b\,{}^0w_b}\right]_{ab}\right\}\right\}$$

Thus our trial function written in full, Eq. (8.31) is

$$\tfrac{1}{2}\mathfrak{D}(\mathfrak{I}_i)=-5kT\left(\tfrac{n_a}{m_a}g_a^{(0)}+\tfrac{n_b}{m_b}g_b^{(0)}\right)-10i(kT)^2B\left(\tfrac{e_an_a}{m_a^3}(g_a^{(0)})^2+\tfrac{e_bn_b}{m_b^3}(g_b^{(0)})^2\right)+$$

$$4(kT)^2\left\{\tfrac{n_a^2}{m_a^2}(g_a^{(0)})^2\left[\overleftrightarrow{w_a\,{}^0w_a},\overleftrightarrow{w_a\,{}^0w_a}\right]_{aa}+\tfrac{n_b^2}{m_b^2}(g_b^{(0)})^2\left[\overleftrightarrow{w_b\,{}^0w_b},\overleftrightarrow{w_b\,{}^0w_b}\right]_{bb}+\right.$$

$$n_an_b\left\{\tfrac{1}{m_a^2}\left[\overleftrightarrow{w_a\,{}^0w_a},\overleftrightarrow{w_a\,{}^0w_a}\right]_{ab}(g_a^{(0)})^2+\tfrac{1}{m_b^2}\left[\overleftrightarrow{w_b\,{}^0w_b},\overleftrightarrow{w_b\,{}^0w_b}\right]_{ab}(g_b^{(0)})^2+\right.$$

$$\left.\left.\tfrac{1}{m_am_b}2g_a^{(0)}g_b^{(0)}\left[\overleftrightarrow{w_a\,{}^0w_a},\overleftrightarrow{w_b\,{}^0w_b}\right]_{ab}\right\}\right\}\tag{8.32}$$

Using the properties of the collision integrals given in Appendix D one finds that the five collision integrals appearing in Eq. (4.4) are related to the collision integral φ defined in Eq. (D.23) (see table in Appendix D) so that Eq. (4.4) when divided by kT yields finally that

$$\frac{\mathfrak{D}(\mathfrak{I}_i)}{2kT}=-5\left(\frac{n_a}{m_a}g_a^{(0)}+\frac{n_b}{m_b}g_b^{(0)}\right)-10ikTB\left(\frac{e_an_a}{m_a^3}(g_a^{(0)})^2+\frac{e_bn_b}{m_b^3}(g_b^{(0)})^2\right)+$$

$$\frac{4kT\varphi}{n}\left[\left\{\sqrt{2}\frac{n_a^2}{m_a^2}(g_a^{(0)})^2+\sqrt{2}M_1\frac{n_b^2}{m_b^2}(g_b^{(0)})^2\right\}+\right.$$

$$\left.\frac{n_an_b}{2}\left\{\frac{2}{m_a^2}g_a^{(0)})^2+\frac{10}{3m_b^2}g_b^{(0)}-\frac{8}{3m_am_b}M_1g_a^{(0)}g_b^{(0)}\right\}\right]\tag{8.33}$$

In Appendix H the main steps outlining the variational procedure to compute $g_a^{(0)}$ and $g_b^{(0)}$ are given as well as the results for all the required coefficients to compute all the quantities involved in the pressure tensor.

Emphasis should be placed upon the fact that since $\alpha_i^{(n)} = \Gamma_i^{(n)}$ (first approximation), the sought values for the Γ's are given through Eqs. (8.15b), (8.16a) and (8.18b) since L_i, G_i and P_i are all related to the solution of Eq. (8.21) so that to first order $G_i \equiv g_i^{(0)}$ and from these follow all the values for the Γ's, or in other words, as we shall see, the values of the five independent viscosities.

8.4 Comparison with Thermodynamics

Eqs. (8.6) and (8.7) have to be shaped into a form allowing for direct comparison with the form predicted, from very general symmetry arguments, for the stress tensor in the presence of a magnetic field. This form is derived in the well known monograph of de Groot & Mazur (see Ref. [5] appendix C) and we shall reproduce it here for the sake of completeness. In our language, remembering that we have taken $\vec{B}$ along the z-axis it takes the following form:

	S_{xx}	S_{yy}	S_{zz}	S_{xz}	S_{yz}	S_{xy}
$(\overleftrightarrow{\tau} - p\mathbb{I})_{xx}$	$-\eta$	$\eta - \eta_3$	0	0	0	η_4
$(\overleftrightarrow{\tau} - p\mathbb{I})_{yy}$	$-\eta + \eta_3$	$-\eta$	0	0	0	η_4
$(\overleftrightarrow{\tau} - p\mathbb{I})_{zz}$	0	0	$-\eta$	0	0	0
$\overleftrightarrow{\tau}_{xy}$	η_4	$-\eta_4$	0	0	0	η_3
$\overleftrightarrow{\tau}_{xz}$	0	0	0	η_2	$-\eta_5$	0
$\overleftrightarrow{\tau}_{yz}$	0	0	0	η_5	η_2	0

(8.34a)

recalling that

$$S_{\alpha\beta} = \frac{1}{2}\left(\frac{\partial u_\alpha}{\partial x_\beta} + \frac{\partial u_\beta}{\partial x_\alpha}\right)\frac{1}{3}\text{div }\vec{u}\delta_{\alpha\beta} \tag{8.34b}$$

and that the bulk viscosity contributions are not present in the dilute gas approximation.

To derive this tensor we notice first in Eq. (8.26a) that the combination $[-B^2(\alpha_i^{(3)} - \alpha_i^{(4)}) + B^4\alpha_i^6]$ vanishes on account of Eq. (8.19). Exactly the same cancellation occurs in Eqs. (8.26c, d) so that in its final form the stress tensor reads,

$$(\overleftrightarrow{\tau} - p\mathbb{I})_{xx} = (kT)^2 \sum_{i=a}^{b} \frac{n_i}{m_i} 2\left\{\Gamma_i^{(1)}\overleftrightarrow{S}_{xx} - \Gamma_i^{(2)} B \overleftrightarrow{S}_{xy} + \Gamma_i^{(3)} B^2 \overleftrightarrow{S}_{yy}\right\} \tag{8.35a}$$

$$(\overleftrightarrow{\tau} - p\mathbb{I})_{yy} = (kT)^2 \sum_{i=a}^{b} \frac{n_i}{m_i} 2\left\{\Gamma_i^{(1)} \overleftrightarrow{S}_{yy} - \Gamma_i^{(2)} B \overleftrightarrow{S}_{xy} + \Gamma_i^{(3)} B^2 \overleftrightarrow{S}_{xx}\right\} \quad (8.35b)$$

$$(\overleftrightarrow{\tau} - p\mathbb{I})_{zz} = (kT)^2 \sum_{i=a}^{b} \frac{n_i}{m_i} 2\Gamma_i^{(1)} \overleftrightarrow{S}_{zz} \quad (8.35c)$$

$$(\overleftrightarrow{\tau})_{xy} = (kT)^2 \sum_{i=a}^{b} \frac{n_i}{m_i} 2\left\{(\Gamma_i^{(1)} + \Gamma_i^{(3)} B^2) \overleftrightarrow{S}_{xy} + \Gamma_i^{(2)} B^2 (\overleftrightarrow{S}_{xx} - \overleftrightarrow{S}_{yy})\right\} \quad (8.35d)$$

$$(\overleftrightarrow{\tau})_{xz} = (kT)^2 \sum_{i=a}^{b} \frac{n_i}{m_i} \left\{2(\Gamma_i^{(1)} + \Gamma_i^{(4)} B^2) \overleftrightarrow{S}_{xz} - B(\Gamma_i^{(2)} + \Gamma_i^{(5)} B^2)(\overleftrightarrow{S}_{yz})\right\} \quad (8.35e)$$

and finally

$$(\overleftrightarrow{\tau})_{yz} = (kT)^2 \sum_{i=a}^{b} \frac{n_i}{m_i} \left\{2(\Gamma_i^{(1)} + \Gamma_i^{(4)} B^2) \overleftrightarrow{S}_{yz} - B(\Gamma_i^{(2)} + \Gamma_i^{(5)} B^2)(\overleftrightarrow{S}_{xz})\right\} \quad (8.34f)$$

One can appreciate the fact that since $\overleftrightarrow{S}$ is a symmetric tensor so is $\overleftrightarrow{\tau}_i$. Further, let us define quantities, all multiplied by $\sum_{i=a}^{b} 2\frac{n_i}{m_i}(kT)^2$ namely

$$\begin{aligned}
\eta &= 2(kT)^2 \textstyle\sum_{i=a}^{b} \frac{n_i}{m_i} \Gamma_i^{(1)} \\
\eta_4 &= 2(kT)^2 \textstyle\sum_{i=a}^{b} \frac{n_i}{m_i} B \Gamma_i^{(2)} \\
\eta - \eta_2 &= 2(kT)^2 \textstyle\sum_{i=a}^{b} \frac{n_i}{m_i} B^2 \Gamma_i^{(4)} \\
\eta - \eta_3 &= 2(kT)^2 \textstyle\sum_{i=a}^{b} \frac{n_i}{m_i} B^2 \Gamma_i^{(3)} \\
\eta_5 &= 2(kT)^2 \textstyle\sum_{i=a}^{b} \frac{n_i}{m_i} B(\Gamma_i^{(2)} + B^2 \Gamma_i^{(5)})
\end{aligned} \quad (8.36)$$

This allows to write the system of Eq. (8.35a-8.34f) in the following way

$$(\overleftrightarrow{\tau} - p\mathbb{I})_{xx} = \vec{v}\eta S_{xx} - \eta_4 \overleftrightarrow{S}_{xy} + (\eta - \eta_3) S_{yy} \quad (8.37a)$$

$$(\overleftrightarrow{\tau} - p\mathbb{I})_{yy} = -\eta \overleftrightarrow{S}_{yy} + \eta_4 \overleftrightarrow{S}_{xy} + (-\eta + \eta_3) \overleftrightarrow{S}_{xx} \quad (8.37b)$$

$$(\overleftrightarrow{\tau} - p\mathbb{I})_{zz} = -\eta \overleftrightarrow{S}_{zz} \tag{8.37c}$$

$$(\overleftrightarrow{\tau})_{xy} = \eta_3 \overleftrightarrow{S}_{xy} + \eta_4(\overleftrightarrow{S}_{xx} - \overleftrightarrow{S}_{yy}) \tag{8.37d}$$

$$(\overleftrightarrow{\tau})_{xz} = \eta_2 \overleftrightarrow{S}_{xz} - \eta_5 \overleftrightarrow{S}_{yz} \tag{8.37e}$$

$$(\overleftrightarrow{\tau})_{yz} = \eta_2 \overleftrightarrow{S}_{yz} + \eta_5 \overleftrightarrow{S}_{xz} \tag{8.37f}$$

From Eqs. (H.1)–(H.4) we immediately see that η_1, η_2 and η_5 are symmetrical under the transformation $\vec{B} \to -\vec{B}$ (even function of $|\vec{B}|$) whereas η_4 and η_5 are odd functions of $|\vec{B}|$ such as required by Onsager's reciprocity theorem. Also $\eta = \eta_3$ when $B = 0$ so all the results correspond to the viscosity of a non-magnetized mixture of dilute gases. With a careful translation of symbols Eqs. (8.36) correspond to the result predicted in LIT as given by Eq. (8.34a).

To understand the behavior of these five viscosities we simply extract from Eqs. (H.1) to (H.4) their real and imaginary parts of $g_i^{(0)}$ where $i = a, b$ to get

$$\begin{aligned} Reg_a^{(0)} &= \tfrac{m_a \tau}{\Delta_1 kT}(6.38 + 0.16x^2) \\ Img_a^{(0)} &= -\tfrac{m_a \tau}{\Delta_1 kT}(24.5x + 0.64x^3) \end{aligned} \tag{8.38}$$

and

$$\begin{aligned} Reg_b^{(0)} &= \tfrac{m_b \tau}{\Delta_1 kT}(0.2386 + 4.091x^2) \\ Img_b^{(0)} &= -\tfrac{m_a \tau}{\Delta_1 kT}(0.037x - 0.6x^3) \end{aligned} \tag{8.39}$$

and

$$\Delta_1 = 6.1 + 104.77\omega_a^2\tau^2 + 2.56\omega_a^4\tau^4 \tag{8.40}$$

The corresponding values for $p_i^{(0)}$ follow readily from the scaling mentioned in p. 99 (c.f. eq. H.2). For instance, for η we have

$$\eta = \frac{n(kT)^2}{m_a}\Gamma_a^{(1)}$$

for a fully ionized plasma, $n_a = n_b = \frac{1}{2}n$.

Using Eq. (H.4) and the value for τ which we rewrite here for convenience,

$$\tau = \frac{1}{\varphi} = \frac{24\pi^{\frac{3}{2}}\sqrt{m_e}\epsilon_0^2}{ne^4\psi}(kT)^{\frac{3}{2}}$$

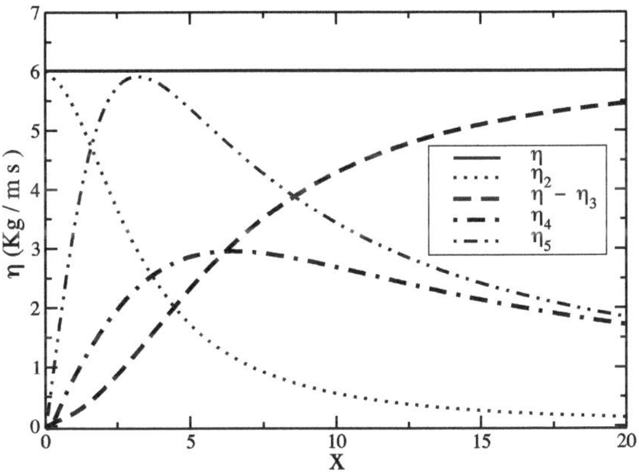

Figure 8.1: The five viscosities as functions of x for $n = 10^{21}$ cm^{-3}, $T = 10^7$ K. Notice the somewhat unexpected result showing that in the presence of a magnetic field additional four viscosities are not negligible when compared with represented by the full line.

we find that

$$\eta = 72.85 \frac{\pi^{\frac{3}{2}} \sqrt{m_e} \epsilon_0^2}{e^4 \psi} (kT)^{\frac{5}{2}} \tag{8.41}$$

where ψ is the logarithmic function defined in Eq. (6.3). This result is in agreement with that obtained both by Balescu and by Braginski. The behavior of the five viscosities is better appreciated in Fig. (8.1), their calculation resulting from simple addition and subtraction of complex numbers. Their relevant characteristic is that they all show the dominating $(kT)^{\frac{5}{2}}$ dependence with temperature. Further, for $\vec{B} \neq \vec{0}$ the values of the viscosities is by no means negligible as compared with η for values of $\omega_e \tau$ field dependent < 10. This result may have strong implications in several situations in astrophysics including the problem of angular momentum transfer in accretion disks in binary systems.

Chapter 9

Magnetohydrodynamics

The subject to be discussed in this chapter is one of the most important and relatively old aspects of fluid dynamics. When a charged fluid is set in motion the charge and current densities generate an electromagnetic field which through Maxwell's equations couples with the fluid's own equations of motion giving rise to what is now called magnetohydrodynamics. As we already pointed out in Chap. 2 the resulting macroscopic equations do not constitute a closed set of equations. Additional information must be supplied through the so-called constitutive equations, relating the different currents or fluxes present to the state variables chosen to describe the fluid. For the particular case of a dilute inert plasma this information was extracted in Chaps. 4–6 from the microscopic model furnished by Boltzmann's kinetic equation when the magnetic field is either homogeneous or weakly inhomogeneous. When the equations are substituted into the conservation equations obtained in Chap. 2 a rather complicated set of non linear partial differential equations will result, the set being complete. Since the calculations have been carried out keeping terms which are linear in the gradients, the thermodynamic forces, we may refer to this set of equations as the Navier-Stokes-Fourier-Maxwell equations of magnetohydrodynamics.

Let us choose for the moment, to describe the non-equilibrium states o a binary plasma, the state variables $\rho_i(\vec{r},t)$ $(i = a, b)$ the local mass densities of each species, $\vec{u}(\vec{r},t)$ the local barycentric velocity and $e(\vec{r},t)$, the total local energy density. Recalling the results from Chap. 2, the equations governing their time dependence are:

$$\frac{\partial \rho_i}{\partial t} + \mathrm{div}(\rho_i \vec{u}) = -\mathrm{div}\ \vec{J}_i \tag{9.1}$$

L.S. García-Colín, L. Dagdug, *The Kinetic Theory of a Dilute Ionized Plasma*, Springer Series on Atomic, Optical, and Plasma Physics 53

the mass conservation equation,

$$\frac{\partial}{\partial t}(\rho\vec{u}) + \text{div}\,(\overleftrightarrow{\tau}^k + \rho\vec{u}\vec{u}) = Q(\vec{E} + (\vec{u}\times\vec{B})) + \vec{J_c}\times\vec{B} \tag{9.2}$$

the momentum conservation equation; and

$$\frac{1}{2}\frac{\partial}{\partial t}(\rho u^2) + \frac{\partial(\rho e)}{\partial t} + \text{div}\,(\vec{J_q} + \vec{u}\cdot\overleftrightarrow{\tau}^k + \rho\vec{u}e + \frac{1}{2}\rho\vec{u}u^2) - \vec{J_T}\cdot\vec{E} = 0 \tag{9.3}$$

the energy equations.

In these equations the mass current $\vec{J_i}$ and the charge conduction current $\vec{J_c}$ are related through,

$$\vec{J_c} = \frac{m_a + m_b}{m_a m_b} e\vec{J_a} \sim \frac{e}{m_a}\vec{J_a} \tag{9.4}$$

the last equality holding when $m_b >> m_a$. Recall that $\vec{J_a} + \vec{J_b} = \vec{0}$. The heat current $\vec{J_q}$ is defined as

$$\vec{J_q} = \sum_i \frac{1}{2}\rho_i\langle\vec{c_i}c_i^2\rangle \tag{9.5}$$

The total charge current is,

$$\vec{J_T} = \sum_i n_i e_i\langle\vec{v_i}\rangle = Q\vec{u} + \vec{J_c} = \frac{m_a + m_b}{m_a m_b}e\vec{J_a} + Q\vec{u} \tag{9.6}$$

since $\vec{v_i} = \vec{u}(\vec{r},t) + \vec{c_i}$ and Q is the total charge.

The momentum current or stress tensor is defined as

$$\overleftrightarrow{\tau}^k = \sum_i \rho_i\langle\vec{c_i}\vec{c_i}\rangle \tag{9.7}$$

and the local fields $\vec{E}(\vec{r},t)$ and $\vec{B}(\vec{r},t)$ are obtained from Maxwell's equations which in the MKS system are,

$$\begin{aligned} &\text{div}\,\vec{E} = \frac{1}{\epsilon_0}Q && \text{div}\,\vec{B} = 0 \\ &\frac{\partial\vec{B}}{\partial t} + \text{rot}\,\vec{E} = 0 && -\frac{\partial\vec{E}}{\partial t} + \text{rot}\,\vec{B} = \mu_0\vec{J_T} \end{aligned} \tag{9.8}$$

We insist on the fact that Eqs. (9.1-9.3) constitute a set of six equations with 21 unknowns. There are six independent variables $(\rho_i, \vec{u}, e)$ and the remaining fifteen constitutive equations, namely, Eqs. (9.5-9.7). The fields $\vec{E}$ and $\vec{B}$ depend also on the two sources Q and $\vec{J_T}$ so that, in principle, the whole time evolution problem of the fluid's state for well specified boundary conditions, is uniquely determined. This constitutes the basic theoretical framework of magnetohydrodynamics at the Navier-Stokes level. The remaining steps required to handle these results are, nevertheless, quite subtle.

Examination of the equations that we have derived for the currents in the previous chapters clearly indicates that their direct substitution into the conservation equations would result in a set of highly non linear partial differential equations so complicated that they are practically unmanageable to deal within a concrete application. But, on the other hand, it is precisely at this stage where in many cases, workers in the field have proceeded in a very arbitrary way.

Let us deal with Eq. (9.2). Let ℓ_H be the characteristic hydrodynamical length, the length of a gradient namely,

$$\ell_H^{-1} = Max \frac{|\text{grad } M(\vec{r})|}{M(\vec{r})}$$

for any local variable $M(\vec{r})$. Let τ_H be a hydrodynamical time, $\tau_H^{-1} = v_s/\ell_H$ where v_s is the velocity of sound. Then, from Maxwell's equations, the ratio

$$\frac{\epsilon_0|\vec{E}\text{div } \vec{E}|}{\frac{1}{\mu_0}|\vec{B}||\vec{J_T} \times \vec{B}|} \sim \epsilon_0\mu_0 \frac{\ell_H^{-1}}{\ell_H^{-1}} v_s \frac{E^2}{B^2} \sim \frac{v_s^2}{c^2}$$

using the fact that $\frac{|\vec{E}|}{\ell_H} \sim \frac{|\vec{B}|}{\tau_H}$. So for a non-relativistic plasma $Q\vec{E}$ is negligible and $(Q\vec{u} + \vec{J_c}) \times \vec{B} = \vec{J_T} \times \vec{B}$. If $\frac{\partial \vec{E}}{\partial t} << 1$, then, by Eq. (9.8)

$$\vec{J_T} \times \vec{B} = \frac{1}{\mu_0}(\text{rot } \vec{B}) \times \vec{B}$$

and Eq. (9.2) will transform into

$$\frac{\partial}{\partial t}(\rho\vec{u}) + \text{div } (\overleftrightarrow{\tau}^k + \rho\vec{u}\vec{u}) = \frac{1}{\mu_0}(\text{rot } \vec{B}) \times \vec{B} \tag{9.9}$$

which is a rather strange result: The charge current, being the source of $\vec{B}$ no longer appears in the equation of motion, a simplification which is debatable. We shall come back to it again.

The energy equation may be also simplified by transforming it into an equation for the internal energy density $e(\vec{r}, t)$. In fact using the continuity equation,

$$\frac{\partial}{\partial t}(\frac{1}{2}\rho u^2) + \text{div}\ \frac{1}{2}\rho u^2 \vec{u} = \rho\vec{u} \cdot \frac{d\vec{u}}{dt}$$

so that multiplying Eq. (9.2) by $\vec{u}$, neglecting the term $Q\vec{E}$ and substituting into (9.3) we get that

$$\rho\frac{de}{dt} + \text{div}\ \vec{J}_q + \overleftrightarrow{\tau}^k : (\text{grad}\ \vec{u})^s + \vec{J}_T \cdot \vec{E}' = 0 \tag{9.10}$$

the balance equation for the internal energy density $e(\vec{r}, t)$. Moreover we already showed that for a mixture, when $m_b \gg m_a$

$$\vec{J}_q = \vec{J}_q' + \frac{5}{2}kT\frac{\vec{J}_a}{m_a}$$

so using Eq. (9.4) and neglecting $Q\vec{E}$, we finally get that

$$\rho\frac{de}{dt} + \text{div}\ \vec{J}_q^{\,\prime} + \overleftrightarrow{\tau}^k : (\text{grad}\ \vec{u})^s + \frac{5k}{2e}\ \text{div}(\vec{J}_c T) + \vec{J}_c \cdot \vec{E}' = 0 \tag{9.11}$$

In order to proceed we now evaluate the contributions arising from the currents in Eqs. (9.2) and (9.11). We recall from Chap. 6 that

$$\vec{J}_c = \frac{n_a e}{m_a}k\left[\alpha_a^{(0)}(\text{grad}\ T)_\parallel + \mathrm{a}_a^{(1)(0)}(\text{grad}\ T)_\perp + \mathrm{a}_a^{(2)(0)}B(\text{grad}\ T)_s\right] +$$
$$\frac{n_a e}{m_a}kT\left[\delta_a^{(0)}(\vec{d}_{ab})_\parallel + d_a^{(1)(0)}(\vec{d}_{ab})_\perp + d_a^{(2)(0)}(d_a B\hat{k} \times \vec{d}_{ab})\right] \tag{9.12}$$

where to simplify things we have chosen the z-*axis* to lie along $\vec{B}$. First notice that if $\vec{B} = \vec{0}$, $\vec{d}_{ab} = -\theta\ \text{grad}\ \phi$ for a fully ionized plasma and $m_b >> m_a$, where $\theta = ne^2/4m_a$. Therefore, with the results of Appendix E, Eq. (9.12) yields,

$$\vec{J}_c = -\sigma_{||}\ \text{grad}\ \phi - \tau_{||}\ \text{grad}\ T \tag{9.13}$$

where

$$\sigma_{||} = \frac{ne^2\tau}{4m_a} \times 1.191 \tag{9.14a}$$

is the electrical conductivity and

$$\tau_{||} = \frac{nke\tau}{4m_a} \times 2.94 \tag{9.14b}$$

is Thomson's thermoelectric coefficient. The first term in Eq. (9.13) is simply the well known Ohm's coefficient. Here arises the first major objection against the approximation leading to Eq. (9.9). These affects are there lost! Eq. (9.9) ignores the possible electrical effects on the plasma even if $\vec{B} = 0$.

If $\vec{B} \neq \vec{0}$, Eq. (9.2) becomes much more complicated. However, for a fully ionized plasma and very small pressure gradients $\vec{d}_{ab}^{(0)}$ may be neglected. Even though, its electromagnetic component together with div $\overleftrightarrow{\tau}^{k'}$ gives a rather awkward expression. The x-component of the equation reads,

$$\begin{aligned}\rho\frac{du_x}{dt} + \frac{\partial\tau_{xx}}{\partial_x} + \frac{\partial\tau_{xy}}{\partial_y} + \frac{\partial\tau_{xz}}{\partial_z} &= (\vec{J}_T \times \vec{B})_x \\ &= Qu_yB + u_yB(J_c)_x\end{aligned}$$

Here,

$$\begin{aligned}(\vec{J}_c)_y = &\frac{n_a ke}{m_a}(\mathrm{a}_a^{(1)(0)}\frac{\partial T}{\partial y} - \mathrm{a}_a^{(2)(0)}B\frac{\partial T}{\partial x})+ \\ &\frac{n_a kTe}{m_a}(-\theta d_a^{(1)(0)}(E_y + u_xB) - \theta d_a^{(2)(0)}(E_x - u_yB))\end{aligned}$$

which finally yields,

$$\begin{aligned}\rho\frac{du_x}{dt}+\frac{\partial\tau_{xx}}{\partial_x}+\frac{\partial\tau_{xy}}{\partial_y}+\frac{\partial\tau_{xz}}{\partial_z} = Bu_x\left(Qu_y - \frac{n_a kTe}{m_a}\left(\frac{\mathrm{a}_a^{(1)(0)}}{T}\frac{\partial T}{\partial y} - \frac{\mathrm{a}_a^{(2)(0)}}{T}\frac{\partial T}{\partial x}\right)\right) \\ -\theta\left(d_a^{(1)(0)}(E_y + u_xB) + d_a^{(2)(0)}(E_x - u_yB)\right) \qquad (9.15)\end{aligned}$$

This is not a very inspiring result. The a_s' and d_s' coefficients are given in Appendix E and are complicated expressions which depend on B, n and T. In the l.h.s the derivatives of the stress tensor are also nasty. We must remember that

$$\begin{aligned}\tau_{xx} &= p + \eta S_{xx} + \eta_4 S_{xy} + (\eta - \eta_3)S_{yy} \\ \tau_{xy} &= \eta_3 S_{xy} + \eta_4(S_{xx} - S_{yy}) \\ \tau_{xz} &= \eta_2 S_{xz} - \eta_5 S_{yz}\end{aligned}$$

and $\overleftrightarrow{S}$ is defined in Eq. (8.26b). Thus the five independent viscosities are involved in the momentum current possibly overshadowing the Navier-Stokes term ηS_{xx}. It is difficult to say anything in general as to their relative importance. Even for an isothermal plasma and vanishing small or inexistent electric fields Eq. (9.15), is impressive,

$$\frac{\partial}{\partial t}(\rho u_x) + (\text{div } \overleftrightarrow{\tau})_x = B\frac{n_a \times k \times T \times e}{m_a}\theta(-u_y d_a^{(1)(0)} + u_x d_a^{(2)(0)})$$

In contrast however, Eq. (9.9) seems to provide a simpler version to the equation of motion since the r.h.s becomes $(1/\mu_0)(\vec{B} \times \text{rot } \vec{B})_x$. But two problems arise. The first one is that the connection with LIT is lost, the second one which requires a deeper thought is how then $\vec{B}$ is related to $\vec{J}_T$? We shall come back to this question later.

Nevertheless, Eq. (9.9) and Eq. (9.15) contain a term, the divergence of the stress tensor which is barely mentioned, if mentioned at all, in the vast literature on the subject of magnetohydrodynamics. In spite of its impressiveness, it is still linear and gives rise to what we could call the extension of the Navier-Newton viscous effects present in an ordinary real fluid. Indeed if one carries out the operations indicated in the term $(\text{div } \overleftrightarrow{\tau})_x$ using the explicit form for the τ_{ij} components and collects terms in a convenient way, the following result is obtained namely,

$$(\text{div } \overleftrightarrow{\tau})_x = \frac{\eta}{2}\left[\frac{\partial^2 u_x}{\partial x^2} - \frac{2}{3}\frac{\partial}{\partial x}\text{div } \vec{u}\right] + \frac{1}{2}\left[\eta\frac{\partial^2 u_x}{\partial x^2} + \eta_3\frac{\partial^2 u_x}{\partial y^2} + \eta_2\frac{\partial^2 u_x}{\partial z^2}\right] +$$
$$\frac{1}{2}\left[\eta_3\frac{\partial^2 u_y}{\partial x \partial y} + \eta_2\frac{\partial^2 u_z}{\partial x \partial z}\right] + (\eta - \eta_3)\left[\frac{\partial^2 u_x}{\partial x \partial y} - \frac{1}{3}\frac{\partial}{\partial x}\text{ div } \vec{u}\right] +$$
$$\eta_4\left[\frac{\partial^2 u_x}{\partial x \partial y} - \frac{\partial^2 u_y}{\partial y^2}\right] - \frac{1}{2}\eta_5\left[\frac{\partial^2 u_y}{\partial z^2} + \frac{\partial^2 u_z}{\partial z \partial y}\right] \tag{9.16a}$$

This equation and the two similar ones for the y and z components respectively, show that the Navier-Newton symmetry characteristic of non-magnetized fluids is completely lost. When $B = 0$, $\eta = \eta_2 = \eta_3$, $\eta_4 = \eta_5 = 0$ and the first three terms reduce to

$$\frac{1}{2}\eta\left(\nabla^2 u_x + \frac{1}{3}\frac{\partial}{\partial x}(\text{div } \vec{u})\right)$$

which is the usual Navier-Newton term if we redefine the viscosity to absorb the 1/2 factor. This is a remarkable result. The presence of a magnetic field is reflected in the "viscous modes" of the plasma through the viscosities η_2, η_3, η_4 and η_5. Further, all terms in Eq. (9.16a) are linear in the velocities so there is no priori reason to ignore them except in those (B, n, T) regimes where such viscosities are unimportant. This is clearly indicated in Fig. 8.1. Just for completeness since they have never been written anywhere, the two other components are:

$$(\text{div } \overleftrightarrow{\tau})_y = \frac{1}{2}\left[\eta_3 \frac{\partial^2 u_y}{\partial x^2} + \eta \frac{\partial^2 u_y}{\partial y^2} + \eta_2 \frac{\partial^2 u_y}{\partial z^2}\right] + \frac{\eta}{2}\left[\frac{\partial^2 u_y}{\partial y^2} - \frac{2}{3}\frac{\partial}{\partial y}\text{div } \vec{u}\right] +$$
$$\frac{1}{2}\left[\eta_3 \frac{\partial^2 u_x}{\partial x \partial y} + \eta_2 \frac{\partial^2 u_z}{\partial z \partial y}\right] + (\eta - \eta_3)\left[\frac{\partial^2 u_x}{\partial x \partial y} - \frac{1}{3}\frac{\partial}{\partial y} \text{ div } \vec{u}\right] +$$
$$\eta_4 \left[\frac{\partial^2 u_x}{\partial x^2} - \frac{2}{3}\frac{\partial^2 u_y}{\partial x \partial y} - -\frac{1}{2}\frac{\partial^2 u_x}{\partial y^2}\right] - \frac{1}{2}\eta_5 \left[\frac{\partial^2 u_y}{\partial z^2} + \frac{\partial^2 u_z}{\partial z \partial y}\right] \tag{9.16b}$$

and

$$(\text{div } \overleftrightarrow{\tau})_z = \left(\frac{1}{2}\eta_2 - \frac{1}{6}\eta\right)\frac{\partial}{\partial z}\text{div } \vec{u} + \frac{1}{2}\left[\eta_2 \frac{\partial^2 u_z}{\partial x^2} + \eta \frac{\partial^2 u_z}{\partial y^2}\right]$$
$$+\frac{\eta}{2}\frac{\partial^2 u_z}{\partial z^2} + \eta_5 \left[\frac{\partial^2 u_x}{\partial x \partial z} - \frac{\partial^2 u_y}{\partial x \partial z}\right] \tag{9.16c}$$

Even in the direction of the magnetic field, the z-axis in our case, the two dimensional Laplacian of the velocity component u_z is affected by the field through η_2. To our knowledge this effect has never been accounted for in elementary magnetohydrodynamics. Recall from Chap. 8 that η_2 is a complicated expression involving the parameter $x = \omega\tau$ and its values are important for x up to 5.

We insist here on the fact that Eqs. (9.16a-c) are all linear in the velocities so that they deserve a much closer attention to find their influence in transport phenomena involving non-relativistic ionized dilute plasmas.

A similar situation arises from the energy equation. Indeed a straightforward application of the local equilibrium assumption transforms Eq. (9.11) into the temperature representation. In fact, since $e = e(\rho, T)$ and noticing that $\left(\frac{\partial e}{\partial \rho}\right)_T = 0$ for an ideal system we get that

$$\rho \frac{de}{dt} = \rho C_v \frac{dT}{dt} = \frac{3}{2} nk \frac{dT}{dt}$$

so Eq. (9.11) reads,

$$\frac{3}{2} nk \frac{dT}{dt} + \text{div } \vec{J}_q^{\,\prime} + \overleftrightarrow{\tau}^k : (\text{grad } \vec{u})^s + \frac{5k}{2e} \text{ div } (\vec{J}_c T) = -\vec{J}_c \cdot \vec{E}' \tag{9.17}$$

Using the expression for $\vec{J}_q{}'$ obtained in Chap. 4, Eq. (4.8) we readily find that its temperature dependent part contributes to div $\vec{J}_q{}'$ by

$$-\kappa_\perp \nabla^2_{(2)} T - \kappa_\| \frac{\partial^2 T}{\partial z^2}$$

and by

$$-\mathbb{D}_\perp \nabla^2_{(2)} p - \mathbb{D}_\| \frac{\partial^2 p}{\partial z^2}$$

where $\nabla^2_{(2)} \equiv \frac{\partial}{\partial x^2} + \frac{\partial}{\partial y^2}$, the κ's and $\mathbb{D}$'s are the parallel and perpendicular expressions for the thermal conductivity and diffusion coefficients, respectively. The contribution arising from $\vec{d}_{ij}$ is readily obtained and reads

$$\text{div } \vec{d}_{ij}^{(e)} = \mathbb{B}_\perp \left(\frac{\partial E_x}{\partial x} + \frac{\partial E_y}{\partial y} \right) + \mathbb{B}_\| \frac{\partial E_z}{\partial z} + \mathbb{B}_\perp \left(\frac{\partial u_y B}{\partial x} + \frac{\partial u_x B}{\partial y} \right) +$$
$$\mathbb{B}_s \left[-\frac{\partial}{\partial x}(E_y + u_x B) + -\frac{\partial}{\partial y}(E_x + u_y B) \right]$$

where the $\mathbb{B}$'s are the Benedicks transport coefficients defined in Chap. 8. Their relative importance in magnetohydrodynamics problem has never been mentioned, even less, assessed. So it is not altogether clear that they are negligible. Even if $\vec{B} = \vec{0}$, and therefore $\mathbb{B}_\| = \mathbb{B}_\perp$ we have that

$$\text{div } \vec{d}_{ij}^{(e)} = \mathbb{B}_\| \nabla^2 \phi$$

a linear contribution in the electrical potential. In astrophysical systems often $\vec{E} = \vec{0}$ and the first two terms disappear. If further, $\vec{B}$ is homogeneous

$$\text{div } \vec{d}_{ij}^{(e)} = B\mathbb{B}_\perp \left(\frac{\partial u_y}{\partial x} + \frac{\partial u_x}{\partial y} \right) - B\mathbb{B}_s \left(\frac{\partial u_x}{\partial x} + \frac{\partial u_y}{\partial y} \right)$$

so its influence depends linearly on the magnitude of the velocity gradients. The contributions arising from the stress tensor are relatively easy to deal with if one is going to neglect non-linear terms. Indeed in the product $\overleftrightarrow{\tau}^k$: (grad $\vec{u}$)^s all contributions are of the type $\tau_{ij} \frac{\partial u_i}{\partial x_j}$, $i,\ j = x,\ y,\ z$ and thus nonlinear. The only term remaining linear in $\vec{u}$ is $-p$div $\vec{u}$ so that this simplifies the temperature equation considerably. Now, the term

$$\frac{5}{2}\frac{k}{e}\text{div } (T\vec{J}_c) = \frac{5}{2}\frac{k}{e}\vec{J}_c \cdot \text{grad } T + \frac{5}{2}\frac{k}{e}T\text{div } \vec{J}_c$$

The first term is clearly a non-linear one in the thermodynamic forces involving grad $T \cdot$ grad T and grad $T \cdot \vec{d}_{ij}$ terms, so we neglect it. Also, since div (grad $T)_s = 0$ the temperature contribution to the Tdiv $\vec{J}_c$ yields,

$$-\frac{n_a k}{m_a} T \left(\alpha_a^{(1)} \frac{\partial^2 T}{\partial z^2} + \mathrm{a}_a^{(0)(1)} \nabla^2 T \right)$$

For a fully ionized plasma, ignoring pressure diffusion effects, we must evaluate the terms arising from $\vec{d}_{ij}^{(e)}$. Indeed

$$\begin{aligned}\text{div } \vec{J}_c^{(e)} =& - \sigma_a^{\|} \frac{\partial E}{\partial z} - \left(\sigma_a^{\perp} \frac{\partial}{\partial x} - \sigma_a^{(s)} \frac{\partial}{\partial y} \right) (E_x + u_y B) + \\ & - \left(\sigma_a^{\perp} \frac{\partial}{\partial x} + \sigma_a^{(s)} \frac{\partial}{\partial y} \right) (E_y - u_x B)\end{aligned}$$

where $\sigma_a^{\|}$, $\sigma_a^{\perp}$ and $\sigma_a^{(s)}$ are defined in Chap. 4. When $\vec{B} = \vec{0}$, $\sigma_a^{(s)} = 0$, $\sigma_a^{\|} = \sigma_a^{\perp}$ and

$$\text{div } \vec{J}_c^{(e)} = -\sigma_a^{\|} \text{div } E$$

Finally,

$$\begin{aligned}\vec{J}_c \cdot \vec{E} =& - \sigma_{\|} E_z^2 - \sigma_{\perp} \left[\left(E^2 - E_z^2 \right) + B \left(u_x E_y - u_y E_x \right) \right] - \\ & \sigma_a^{(s)} B \left[\left(u_y E_y - u_x E_x \right) \right] \\ =& - \sigma_{\|} E^2 \qquad \text{if} \quad B = 0\end{aligned}$$

the well known Ohmic dissipation term.

Inserting all these results in Eq. (9.17) the temperature equation, we obtain that,

$$\begin{aligned}& \frac{3}{2} nk \frac{dT}{dt} - \kappa_{\perp} \nabla_{(2)}^2 T - \kappa_{\|} \frac{\partial^2 T}{\partial z^2} - \mathbb{D}_{\perp} \nabla_{(2)}^2 p - \mathbb{D}_{\|} \frac{\partial^2 p}{\partial z^2} + \\ & \mathbb{B}_{\perp} \left(\frac{\partial E_x}{\partial x} + \frac{\partial E_y}{\partial y} \right) + \mathbb{B}_{\|} \frac{\partial E_z}{\partial z} + \mathbb{B}_{\perp} \left(\frac{\partial u_y B}{\partial x} + \frac{\partial u_x B}{\partial y} \right) + \\ & \mathbb{B}_s \left[- \frac{\partial}{\partial x} (E_y + u_x B) + \frac{\partial}{\partial y} (E_x + u_y B) \right] \\ & - \sigma_{\|} E_z^2 - \sigma_{\perp} \left[\left(E^2 - E_z^2 \right) + B \left(u_x E_y - u_y E_x \right) \right] - \\ & \sigma_a^{(s)} B \left[\left(u_y E_y - u_x E_x \right) \right] - \frac{n_a k}{m_a} T \left(\alpha_a^{(1)} \frac{\partial^2 T}{\partial z^2} + \mathrm{a}_a^{(0)(1)} \nabla_{(2)}^2 T \right) \\ & + p \text{ div } \vec{u} = 0\end{aligned} \tag{9.18}$$

Equations (9.1), (9.15), and (9.18) are the linear approximation to the full set of the equations of magnetohydrodynamics to first order in the gradients for a fully ionized hydrogen-like plasma ($m_b >> m_a$). Even with the approximations introduced they are still quite unmanageable and also far from clear how to devise a suitable systematic way to obtain appropriate approximations. Thus, as already expressed by Balescu [4], for any specific problem one will have to find the way of reducing this set into one which is simpler to operate with.

It is pertinent at this point to discuss a different approach to the interpretation of the energy equation similar to that leading to Eq. (9.9) for the momentum equation. Indeed, in Eq. (9.17) $\vec{J}_c \cdot \vec{E}'$ may be substituted by $\vec{J}_T \cdot \vec{E}'$ and assuming that $\partial_t \vec{E} \ll 1$, $\vec{J}_T = \mu_0^{-1} \text{rot } \vec{B}$ so that

$$\vec{J}_T \cdot \vec{E}' = \frac{1}{\mu_0} \text{rot } \vec{B} \left(\vec{E} + \vec{u} \times \vec{B} \right)$$

Now we introduce a second assumption, namely that the constitutive equation for $\vec{J}_c$ given by

$$\vec{J}_c = \sigma_{\parallel} E_z \hat{k} + \sigma_{\perp} \left(\hat{\imath} E_x + \hat{\jmath} E_y \right) + \sigma_s \left(\vec{E} + \vec{u} \vec{B} \right) \tag{9.19a}$$

ignoring pressure diffusion and thermoelectric effects as inferred from Eq. (9.12). This equation is invariably substituted by

$$\vec{J}_T = \sigma_{\parallel} \left(\vec{E} + \vec{u} \vec{B} \right) \tag{9.19b}$$

which is difficult to understand. Indeed $\sigma_{\perp} < \sigma_{\parallel}$ is overrated in the first two terms by assuming equality and σ_s completely misunderstood in evaluating the third term, where $\vec{E}$ is neglected and $\sigma_s = \sigma_{\parallel}$. Nevertheless, from this result

$$\vec{E} + \vec{u} \times \vec{B} = \frac{1}{\mu_0 \sigma_{\parallel}} (\text{rot } \vec{B})$$

and therefore,

$$\vec{J}_T \cdot \vec{E}' = \frac{1}{\mu_0^2 \sigma_{\parallel}} (\text{rot } \vec{B})^2 \tag{9.19}$$

The temperature equation is completely modified since also

$$\text{div } \vec{J}_c = \text{div } \vec{J}_T = \text{div } (\text{rot } \vec{B}) = 0 \tag{9.20}$$

whence

$$\frac{3}{2}nk\frac{dT}{dt} - \kappa_\perp \nabla^2_{(2)} T - \kappa_\| \frac{\partial^2 T}{\partial z^2} + p \text{ div } \vec{u} = -\frac{1}{\mu_0^2 \sigma_\|}(\text{rot } \vec{B})^2 \tag{9.21}$$

ignoring pressure diffusion.

Setting aside the fact that these results are hard to believe less to understand, they are in complete contradiction with the tenets of LIT. $\vec{J}_c$, the conduction current is the outcome of the presence of all thermodynamic forces present in the system. This means that the direct electromagnetic effects represented in the second term of Eq. (9.12) are accompanied by the cross effects arising from diffusive forces $\vec{d}_{ij}^{(0)}$ and temperature gradients. Some may predominate over the others but clearly having all acting in a way such that div $\vec{J}_c = 0$ as required by Eq. (9.20) is simply untenable. Indeed even in the absence of a magnetic field,

$$\text{div } \vec{J}_c = -\sigma_\| \nabla^2 \phi - \tau_{11} \nabla^2 T$$

which could only be zero if grad $\phi = -\vec{E}$ and grad T acted in a rather peculiar way. We heartfully believe that this approach to magnetohydrodynamics is wrong and that the correct form for the temperature equation is that given by Eq. (9.17).

In order to get a better assessment of the results here derived, it is important at this stage to seek for their relationship with the Euler equations, also named ideal or non-resistive equations, of magnetohydrodynamics. They were derived in Chap. 3 but we repeat them here keeping the approximation that $Q|\vec{E}|$ is negligible. The continuity equation is the same, the other two are

$$\rho\frac{\partial \vec{u}}{\partial t} = -\rho \vec{u} \cdot \text{grad } \vec{u} - \text{grad } p + \vec{J}_T \times \vec{B} \tag{3.7}$$

and

$$\frac{\partial T}{\partial t} = -\left(\vec{u} \cdot \text{grad } T + \frac{2p}{3nk} \text{ div } \vec{u}\right) \tag{3.8}$$

were, for the time being, we restrain from using the questionable result asserting that $\vec{J}_T = \mu_0^{-1}\text{rot } \vec{B}$. Further, $\vec{J}_T \times \vec{B} = Q\vec{u} \times \vec{B} + \vec{J}_c \times \vec{B}$ and this second term follows readily from Eq. (9.12) indicating clearly the source terms still present at the Euler level. Yet, in this canonical form they are never used in the literature.

On the other hand, let us go back to Eq. (3.7) and keep all terms. Using the first of Maxwell' equations

$$\rho\frac{\partial \vec{u}}{\partial t} = -\rho\vec{u}\cdot\text{grad }\vec{u} - \text{grad }p + \epsilon_0(\text{div }\vec{E})\vec{E} + \vec{J}_T\times\vec{B}$$

Using the fourth of Maxwell' equations,

$$\vec{J}_T\times\vec{B} = \frac{1}{\mu_0}\left(\partial_t\vec{E} + \text{rot }\vec{B}\right)\times\vec{B}$$

where $\partial_t \equiv \partial/\partial t$, the step being completely legitimate. Now, using once more Maxwell' equations

$$\frac{1}{\mu_0}\left(\partial_t\vec{E} + \text{rot }\vec{B}\right)\times\vec{B} = \frac{1}{\mu_0}\frac{\partial}{\partial t}\left(\vec{E}\times\vec{B}\right) - \frac{1}{\mu_0}\vec{B}\times\text{rot }\vec{E} + \frac{1}{\mu_0}\left(\text{rot }\vec{B}\right)\times\vec{B}$$

But,

$$\left(\text{rot }\vec{B}\right)\times\vec{B} = \text{ div }\left(\vec{B}\vec{B}\right) - \left(\vec{B}\cdot\text{grad }\right)\vec{B}$$

and

$$\vec{E}\text{ div }\vec{E} = -\left(\vec{E}\cdot\text{grad }\vec{E}\right) + \text{ div }\left(\vec{E}\vec{E}\right)$$

Therefore,

$$\epsilon_0\vec{E}\text{ div }\vec{E} + \vec{J}_T\times\vec{B} = -\,\epsilon_0\left(\vec{E}\cdot\text{grad }\vec{E}\right) + \epsilon_0\text{ div }\left(\vec{E}\vec{E}\right) - \frac{1}{\mu_0}\partial_t\left(\vec{E}\times\vec{B}\right) - \frac{1}{\mu_0}\vec{B}\text{grad }\vec{B} - \frac{1}{\mu_0}\text{ div }(\vec{B}\vec{B})$$

Defining

$$\vec{G} = \frac{1}{\mu_0}\left(\vec{E}\times\vec{B}\right) \tag{9.22}$$

and

$$\overleftrightarrow{\tau}^e = \left[\left(\frac{\epsilon_0}{2}|\vec{E}|^2 + \frac{1}{2\mu_0}|\vec{B}|^2\right)\mathbb{I} - \left(\epsilon_0\vec{E}\vec{E} + \frac{1}{\mu_0}\vec{B}\vec{B}\right)\right] \tag{9.23}$$

the electromagnetic stress tensor, and introducing these expressions in Eq. (3.7) we get,

$$\rho\frac{\partial \vec{u}}{\partial t} = -\rho\vec{u}\cdot\text{grad }\vec{u} - \text{div }(p\mathbb{I}) - \frac{\partial_t G}{\partial t} + \text{div }\overleftrightarrow{\tau}^e = 0 \tag{9.24}$$

Eq. (9.24) is the complete Euler equation for an ionized plasma. It shows that even without explicitly writing the term $\vec{J}_T \times \vec{B}$ it is completely equivalent to Eq. (3.7). If we call $\vec{M} \equiv \rho\vec{u}$ the mechanical momentum, Eq. (9.24) reads after integration over an arbitrary volume V,

$$\frac{\partial}{\partial t}\int_V \left(\vec{M}+\vec{G}\right) dV + \int_S \left(\overleftrightarrow{\tau} + \overleftrightarrow{\tau}^e\right)\cdot d\vec{\sigma} \tag{9.25}$$

where $\overleftrightarrow{\tau} = p\mathbb{I}$, no dissipative terms are present. In words: The rate of increase (or decrease) of the momentum of particles and fields equals the rate at which momentum is following out of (or into) an arbitrary volume V.

As for the energy equation (3.8) it is self explanatory so that together with (3.7) and the continuity equation may be regarded as the set of Euler equations for a dilute plasma derived from the microscopic model furnished by the Boltzmann equation.

For completeness, it should be mentioned that the conservation theorem expressed in Eq. (9.25) follows readily from the full momentum equation, Eq. (9.25) provided the mechanical stress tensor $\overleftrightarrow{\tau}$ is now considered as the full tensor $\overleftrightarrow{\tau} = p\mathbb{I} + \overleftrightarrow{\tau}^k$. The interested reader may easily verify this statement.

We would like to finish this chapter on magnetohydrodynamics by discussing what most authors in the subject accept as the equations of "resistive magnetohydrodynamics" emphasizing the complete lack of support they have from the Boltzmann microscopic model of a plasma. We start from Eq. (9.12) which may be regarded as the canonical form for the constitutive equations for the conduction current to first order in the gradients. If in this equation we keep only the electromagnetic terms arising from the electric component of $\vec{d}_{ij}$, which means ignoring thermoelectricity, pressure diffusion and so on, we get Eq. (9.19a), where the three electrical conductivities are thoroughly discussed in Chap. 6. This equation, as mentioned earlier, is arbitrarily substituted by Eq. (9.19b). In Eq. (9.19b) $\sigma_\perp$ is notoriously overrated setting it equal to $\sigma_\parallel$ which is only true when $\vec{B}$ vanishes; the term $\sigma_s\vec{E}$ is ignored and worst σ_s is set equal to $\sigma_\parallel$. All these assumptions are completely wrong since $\sigma_s < \sigma_\perp < \sigma_\parallel$ for all values of $B \neq 0$. Nevertheless they allow writing that

$$\vec{E} = \frac{1}{\sigma_\parallel}\vec{J}_T - \left(\vec{u}\times\vec{B}\right)$$

Assuming that $\partial_t\vec{E} \ll 1$, an assumption which destroys the symmetry behind the argument leading to Eq. (9.25), one may set $\vec{J}_T = \mu_0^{-1}\mathrm{rot}\ \vec{B}$ therefore leading to

$$\vec{E} + \vec{u} \times \vec{B} = \frac{1}{\mu_0 \sigma_\parallel} \text{rot } \vec{B} \tag{9.26}$$

Even worst, from the first of Maxwell's equations and Eq. (9.26)

$$Q = -\frac{1}{\epsilon_0} \text{ div } \vec{u} \times \vec{B}$$

the two sources of the fields turn to depend only on the fields themselves. This is preposterous. In fact from Maxwell's equations,

$$\frac{\partial \vec{B}}{\partial t} = -\text{rot } \vec{E} = \text{rot } \left(\vec{u} \times \vec{B}\right) - \frac{1}{\mu_0 \sigma_\parallel} (\text{rot }) \left(\text{rot } \vec{B}\right)$$

But $\left(\text{rot }\right)\left(\text{rot } \vec{B}\right) = \text{grad } (\text{div } \vec{B}) - \nabla^2 \vec{B}$ whence

$$\frac{\partial \vec{B}}{\partial t} = \text{rot } \left(\vec{u} \times \vec{B}\right) + \frac{1}{\mu_0 \sigma_\parallel} \nabla^2 \vec{B} \tag{9.27}$$

implying that the magnetic field is autonomous but still has to comply with the condition that rot $\vec{B} = \mu_0 \vec{J}_T$!!

Moreover, the equation of motion which in the Euler's regime is given by Eq. (3.7), it is now transformed into

$$\rho \frac{\partial \vec{u}}{\partial t} = -\text{grad } p - \frac{1}{\mu_0} \vec{B} \times \text{rot } \vec{B} \tag{9.28}$$

thus ignoring all dissipative effects coming from div $\overleftrightarrow{T}^k$.

Finally the equation for the pressure is obtained from Eq. (9.18) neglecting all thermoelectric and Benedick's effects, $\kappa_{\parallel,\perp} = 0$, $\mathbb{B}_{\parallel,\perp} = 0$; all diffusive terms setting $\sigma_\perp B(u_x E_y - u_y E_x)$=0, $\sigma_s = 0$ and letting $\sigma_\parallel = \sigma_\perp$ so that

$$\frac{3}{2} nk \frac{dT}{dt} + p \text{ div } \vec{u} = \sigma_\parallel E^2 \tag{9.29}$$

Using the equation of state, Dalton's law

$$p = nkT \quad \text{if} \quad m_b >> m_a$$

so that

$$\frac{dp}{dt} = kT \frac{dn}{dt} + nk \frac{dT}{dt} = \frac{p}{n} \frac{dn}{dt} + \frac{p}{T} \frac{dT}{dt}$$

we obtain that, so

$$nk\frac{dT}{dt} = \frac{dp}{dT} - \frac{p}{\rho}\frac{d\rho}{dt}$$

With the continuity equation this result is transformed into,

$$nk\frac{dT}{dt} = \frac{dp}{dt} + p\text{div }\vec{u}$$

so that equation (9.29) reads,

$$\frac{dp}{dt} + \frac{5}{3}p\text{div }\vec{u} = \frac{2}{3}\frac{1}{\mu_0\sigma_{\|}^2}\text{rot }\vec{B}\cdot\text{rot }\vec{B} \tag{9.30}$$

Eqs. (9.27), (9.28) and (9.30) are referred to in the literature as the equations for "resistive magnetohydrodynamics". We believe that there is not much to add about them. Within the framework provided by kinetic theory they constitute a rather arbitrary model whose implications can be seriously considered suspect. They are completely unjustified. It is clear that according to this work, the correct equations for dissipative magnetohydrodynamics, to first order in the gradients are provided by Eqs. (9.1), (9.15) and (9.18). Their application requires we insist, a very careful examination of the problem to be dealt with before any kind of simplifications may be imposed on them. This is the great and deep value of kinetic theory.

Bibliography

[1] *Plasma Physics* by S. Chandrasekhar. The University of Chicago Press, Chicago (1960).

[2] J. L. Delcroix; *Introduction to the Theory of Ionized Gases*; Interscience Publishers Inc., New York (1960).

[3] R. M. Kulsrud; *Plasma Physics for Astrophysics*; Princeton Univ. Press, Princeton, N. J. (2005), Chap. 3 and references there in.

[4] R. Balescu; *Transport Processes in Plasmas*; Vol. 1. *Classical Transport*; North-Holland Publishing Co., Amsterdam (1988).

[5] L. Spitzer Jr.; *The Physics of Fully Ionized Gases*; Wiley-Interscience, New York (1962).

Appendix A

Calculation of M

We start by substituting Eq. (3.15') in the first two terms of Eq. (3.16a) to get

$$-f_a^{(0)}\frac{e_a}{m_a}(\vec{c}_a \times \vec{B}) \cdot \left(\frac{\partial}{\partial \vec{c}_a}\vec{c}_a \mathbb{A}_a^{(1)} + (\vec{c}_a \times \vec{B})\mathbb{A}_a^{(2)} + \vec{B}(\vec{c}_a \cdot \vec{B})\mathbb{A}_a^{(3)}\right)$$

$$-f_a^{(0)}\frac{m_a}{\rho kT}\left\{\sum_j e_j \int d\vec{c}_j f_j^{(0)} \mathbb{A}_j(\vec{c}_j \times \vec{B})\right\}\vec{c}_a \equiv M$$

and examine term by term.

$$\text{a)}: \quad (\vec{c}_a \times \vec{B}) \cdot \frac{\partial}{\partial \vec{c}_a}(\vec{c}_a \mathbb{A}_a^{(1)}) - (\vec{c}_a \times \vec{B})\mathbb{A}_a^{(1)} + (\vec{c}_a \times \vec{B}) \cdot \vec{c}_a \frac{\partial \mathbb{A}_a^{(1)}}{\partial \vec{c}_a}$$

The second term vanishes so we keep only the first one

$$\text{b)}: \quad (\vec{c}_a \times \vec{B}) \cdot \left[\frac{\partial}{\partial \vec{c}_a}(\vec{c}_a \times \vec{B})\mathbb{A}_a^{(2)}\right] = (\vec{c}_a \times \vec{B}) \cdot \left[\frac{\partial}{\partial \vec{c}_a}(\vec{c}_a \times \vec{B})\right]\mathbb{A}_a^{(2)} +$$

$$(\vec{c}_a \times \vec{B}) \cdot \left[(\vec{c}_a \times \vec{B})\frac{\partial \mathbb{A}_a^{(2)}}{\partial \vec{c}_a}\right]$$

The second term vanishes since $\frac{\partial \mathbb{A}_a^{(2)}}{\partial \vec{c}_a} \sim \vec{c}_a$ and the first term we compute using the vector identity,

$$\text{grad}\left[(\vec{c}_a \times \vec{B}) \cdot (\vec{c}_a \times \vec{B})\right] = 2(\vec{c}_a \times \vec{B}) \cdot \text{grad}(\vec{c}_a \times \vec{B}) + 2(\vec{c}_a \times \vec{B}) \times \text{rot}(\vec{c}_a \times \vec{B})$$

L.S. García-Colín, L. Dagdug, *The Kinetic Theory of a Dilute Ionized Plasma*,
Springer Series on Atomic, Optical, and Plasma Physics 53

where grad and rot are taken with respect to $\vec{c}_a$. Using well known vector identities, we get

$$2(\vec{c}_a \times \vec{B}) \cdot \text{grad}(\vec{c}_a \times \vec{B}) = \text{grad}\left(c_a^2 B^2 - (\vec{c}_a \cdot \vec{B})^2\right) -$$

$$2(\vec{c}_a \times \vec{B}) \times \left[\vec{c}_a \cdot \text{grad}B - \vec{B} \text{ grad } \vec{c}_a + \vec{c}_a \text{ div } \vec{B} - \vec{B} \text{ div } \vec{c}_a\right]$$

and all second term vanishes, except $-\vec{B}$ grad $\vec{c}_a = -\vec{B}$, since $\vec{B}$ is constant in $\vec{c}_a$ space and $(\vec{c}_a \times \vec{B}) \times \vec{B} = (\vec{c}_a \cdot \vec{B})\vec{B} - B^2\vec{c}_a$. But $\text{grad}(c_a^2 B^2 - (\vec{c}_a \cdot \vec{B})^2) = 2B^2\vec{c}_a(\vec{c}_a \cdot \vec{B})\vec{B}$ so adding the two terms we get that

$$(\vec{c}_a \times \vec{B}) \cdot \text{ grad } (\vec{c}_a \times \vec{B}) = 2(B^2\vec{c}_a - \vec{B}(\vec{c}_a \cdot \vec{B}))$$

c): The third term vanishes since $(\vec{c}_a \times \vec{B}) \cdot \vec{c}_a = 0$.
Thus the first term in M reads,

$$-f_a^{(0)} \frac{e_a}{m_a} (\vec{c}_a \times \vec{B}) \mathbb{A}_a^{(1)} - f_a^{(0)} \frac{2e_a}{m_a} \left((B^2\vec{c}_a - \vec{B}(\vec{c}_a \cdot \vec{B}))\right) \mathbb{A}_a^{(2)} \tag{A.1}$$

We first simplify the second term in M to read

$$-f_a^{(0)} \frac{m_a}{\rho kT} \left\{ \sum_j e_j \int d\vec{c}_j f_j^{(0)} \mathbb{A}_j \vec{c}_j \cdot (\vec{c}_a \times \vec{B}) \right\} =$$

$$f_a^{(0)} \frac{m_a}{\rho kT} (\vec{B} \times \vec{c}_a) \cdot \left\{ \sum_j e_j \int d\vec{c}_j f_j^{(0)} \vec{c}_j \left(\mathbb{A}_j^{(1)} \vec{c}_j + (\vec{c}_j \times \vec{B}) \mathbb{A}_j^{(2)} + \vec{B}(\vec{c}_j \cdot \vec{B}) \mathbb{A}_j^{(3)} \right) \right\}$$

Again, proceed term by term,

$$\sum_j e_j \int d\vec{c}_j f_j^{(0)} \mathbb{A}_j^{(1)} \vec{c}_j \vec{c}_j = \frac{1}{3} \sum_j \int d\vec{c}_j f_j^{(0)} \mathbb{A}_j^{(1)} c_j^2 \mathbb{I} \tag{A.2}$$

The second term is

$$\sum_j e_j \int d\vec{c}_j f_j^{(0)} \mathbb{A}_j^{(2)} (\vec{c}_j \times \vec{B}) \vec{c}_j = \sum_j e_j \int d\vec{c}_j f_j^{(0)} \mathbb{A}_j^{(2)} \vec{c}_j \vec{c}_j \times \vec{B}$$

$$= \left(\frac{1}{3} \sum_j e_j \int d\vec{c}_j f_j^{(0)} \mathbb{A}_j^{(2)} c_j^2 \mathbb{I} \right) \times \vec{B}$$

However

$$\left[(\vec{c}_a \times \vec{B}) \cdot \mathbb{I}\right] \times \vec{B} = (\vec{c}_a \times \vec{B}) \times \vec{B}$$

and this term yields

$$\left[B^2\vec{c}_a - (\vec{c}_j \cdot \vec{B})\vec{B}\right] \frac{1}{3} \sum_j e_j \int d\vec{c}_j f_j^{(0)} \mathbb{A}_j^{(2)} c_j^2 \tag{A.3}$$

The third term vanishes since $(\vec{c}_a \times \vec{B}) \cdot \vec{B} = 0$

Using the identity

$$\int \left[c_j^2 - \frac{1}{B^2}(\vec{c}_a \cdot \vec{B})^2\right] f_j^{(0)} R(c) d\vec{c} = \frac{2}{3} \int d\vec{c}_j f_j^{(0)} R(c) c_j^2$$

which is valid for any arbitrary function of c, $R(c)$, we finally find combining Eqs. (A.2-A.3) that

$$M = -f_a^{(0)} \frac{e_a}{m_a} (\vec{c}_a \times \vec{B}) \mathbb{A}_a^{(1)} - f_a^{(0)} \frac{2e_a}{m_a} \left(B^2\vec{c}_a - \vec{B}(\vec{c}_a \cdot \vec{B})\right) \mathbb{A}_a^{(2)} +$$

$$f_a^{(0)} \frac{m_a}{\rho kT} (\vec{c}_a \times \vec{B}) \mathbb{G}_B^{(1)} + f_a^{(0)} \frac{m_a}{\rho kT} \left(B^2\vec{c}_a - \vec{B}(\vec{c}_a \cdot \vec{B})\right) \mathbb{G}_B^{(2)}$$

which are the results quoted in the text. $\mathbb{G}_B^{(1)}$ and $\mathbb{G}_B^{(2)}$ are defined in the text in page 31.

Appendix B

Linearized Boltzmann Collision Kernels

In what follows we shall discuss several important properties related to the nature and structure of the linearized Boltzmann collision kernels. Let $G_{ij} = G_i(\vec{\omega}_i, \vec{\omega}_j)$ and $H_{ij} = H_{ij}(\vec{\omega}_i, \vec{\omega}_j)$ be any two tensors functions of the velocities $\vec{\omega}_i$ and $\vec{\omega}_j$. Define

$$[G_{ij}, H_{ij}]_{ij} \equiv \frac{1}{n_i n_j} \int \cdots \int G_{ij} : (H'_{ij} - H_{ij}) f_i^{(0)} f_j^{(0)} g_{ij} \sigma(\Omega) d\Omega d\vec{v}_i d\vec{v}_j \quad \text{(B.1)}$$

where the subscript in the bracket denotes an integration over the variables $\vec{v}_i$ and $\vec{v}_j$ and the differential cross section $\sigma(\Omega)d\Omega = bdbd\epsilon$ where ϵ is the inclination of the orbit and b the impact parameter. Setting $\vec{v}_i \to \vec{v'}_i$, $\vec{v}_j \to \vec{v'}_j$ noticing that $g_{ij} = g'_{ij}$, $f_i^{(0)}(\vec{v}_i) f_j^{(0)}(\vec{v}_j) = f_i^{(0)}(\vec{v'}_i) f_j^{(0)}(\vec{v'}_j)$ and changing signs in (B.1) we get that

$$[G_{ij}, H_{ij}]_{ij} \equiv -\frac{1}{2n_i n_j} \int \cdots \int (G'_{ij} - G_{ij}) : (H'_{ij} - H_{ij}) f_i^{(0)} f_j^{(0)} g_{ij} \sigma(\Omega) d\Omega d\vec{v}_i d\vec{v}_j \quad \text{(B.2)}$$

The bracket is symmetrical with respect to an exchange of the indexes i and j in G_{ij} and H_{ij} and also between brackets. Hence the following set of equations holds:

$$[G_{ij}, H_{ij}]_{ij} = [H_{ij}, G_{ij}]_{ij} = [G_{ij}, H_{ij}]_{ji} = [H_{ij}, G_{ij}]_{ji} \quad \text{(B.3)}$$

It is also clear from its definition that $[\quad]_{ij}$ is a linear operator. Suppose that

$$G_{ij} = K_i + L_j \qquad H_{ij} = M_i + N_j$$

L.S. García-Colín, L. Dagdug, *The Kinetic Theory of a Dilute Ionized Plasma*, Springer Series on Atomic, Optical, and Plasma Physics 53

K_i and L_i depend only on $\vec{\omega}_i$ and L_j, N_j depend only on $\vec{\omega}_j$, then by inspection,

$$\begin{aligned}[K_i + L_j, M_i + N_j]_{ij} &= [K_i, M_i + N_j]_{ij} + [L_j, M_i + N_j]_{ij} \\ &= [K_i, M_i]_{ij} + [K_i, N_j]_{ij} + [L_j, M_i]_{ij} + [L_j, N_j]_{ij} \end{aligned} \tag{B.4}$$

In Eq. (B.4), $K_i \equiv K_i(\vec{\omega}_i)$, $L_j \equiv L_j(\vec{\omega}_j)$ and so on, and $[\quad]_{ij}$ indicates that the integral is evaluated for collisions between molecules of species i and j. Let us now consider two sets of tensor functions K_i and L_i and define

$$\{K; L\} = \sum_{ij} n_i n_j \left[K_i + K_j; L_i + L_j\right]_{ij} \tag{B.5}$$

From the properties of the square bracket it follows immediately that

$$\{K, L\} = \{L, K\} \tag{B.6a}$$

and

$$\{K; L + M\} = \{K, L\} + \{K, M\} \tag{B.6b}$$

Since $\{K; K\}$ represents integrals whose integrands are non-negative, it follows that

$$\{K, K\} \geq 0 \tag{B.7}$$

and further, for obvious reasons, the equality sign holds if K is a linear combination of the collisional invariants.

For a binary mixture, expanding (B.5) we get that

$$\{K, L\} = n_a^2 \left[K_a + K_a, L_a + L_a\right]_{aa} + 2 n_a n_b \left[K_a + K_b, L_a + L_b\right]_{ab} +$$
$$n_b^2 \left[K_b + K_b, L_b + L_b\right]_{bb}$$

which upon expansion and appropriate collection of terms yields

$$\{K, L\} = 2 n_a^2 \left[K_a, L_a\right]_{aa} + 2 n_b^2 \left[K_b, L_b\right]_{bb} +$$
$$2 n_a n_b \left([K_a, L_a]_{ab} + [K_a, L_b]_{ab} + [K_b, L_a]_{ab} + [K_b, L_b]_{ab}\right) \tag{B.8}$$

Eqs. (B.7) and (B.8) are of importance in the solution to the integral equations.

Bibliography

[1] J. O. Hirschfelder, C. F. Curtiss and R. B. Byrd; *The Molecular Theory of Liquids and Gases*; John Wiley & Sons, New York (1964), $2^{\underline{nd}}$ printing.

Appendix C

The Case when $\vec{B} = \vec{0}$

There are at least two reasons to take the time and space to consider this case in some detail. Firstly and above all, the fact that the full thermodynamic theory can be derived including the explicit form of all transport for the mixture, and a rigorous proof showing that the transport matrix is symmetric in full agreement with Onsager's reciprocity theorem. Secondly, the explicit expressions for the thermal and electrical conductivities can be compared with those derived earlier by Spitzer [1]. Moreover the cross coefficients for the Soret and Dufour effects are also readily obtained which to the author's knowledge have never been appropriately assessed in the case of a ionized gas. Their values could be significant in some astrophysical systems such as cooling flows and planetary nebulae.

Many of the results to be given here arise simply from those in the main text just setting $\vec{B} = 0$. Others require some additional attention which will be offered in detail. The conservation equations are given by Eq. (2.17c) which remains unchanged. The momentum equation is

$$\frac{\partial}{\partial t}(\rho\vec{u}) + \text{div}\,(\tau^k + \rho\vec{u}\vec{u}) = Q\vec{E} \tag{C.1}$$

which may be readily derived [c.f. Eq. (2.9)] and from Eq. (2.19)

$$\rho\frac{d}{dt}\left(\frac{e}{\rho}\right) + \text{div}\,\vec{J}_q + \tau_k : \text{grad}\,\vec{u} = \vec{J}_c \cdot \vec{E} \tag{C.2}$$

Nothing else changes with respect to the H-theorem, the validity of Eq. (2.26) giving the equilibrium distribution function nor the fact that the solution to the linearized homogeneous term of the Boltzmann equation $J(f^0 f^0) = 0$

L.S. García-Colín, L. Dagdug, *The Kinetic Theory of a Dilute Ionized Plasma*, Springer Series on Atomic, Optical, and Plasma Physics 53

is the local Maxwell-Boltzmann distribution function. The first substantial difference with the $\vec{B} \neq 0$ case is Eq. (2.29a) for the first order in gradients solution to the BE, namely,

$$\frac{\partial f_a^{(0)}}{\partial t} + \vec{v_a} \cdot \frac{\partial f_i^{(0)}}{\partial \vec{r}} + \frac{\vec{F_a}}{m_a} \cdot \frac{\partial f_a^{(0)}}{\partial \vec{v_a}} = f_a^{(0)} \left\{ C(\varphi_a^{(1)} + C(\varphi_a^{(1)} + \varphi_b^{(1)} \right\}$$

where $\vec{F_a} = \vec{F_a}^{(e)} + e_a\vec{E}$. Since both $\vec{F_a}^{(e)}$ and $e_a\vec{E}$ are conservative forces it may be readily identified with the linearized version of the Boltzmann equation in the traditional case, $\vec{F}$ conservative. Evaluation of the left hand side using the explicit form for $f^{(0)}$ and Eqs. (3.5)-(3.8) in the text where the last two terms in (3.7) are replaced by $Q\vec{E}$ leads to the result that

$$f_a^{(0)} \left\{ \frac{m_a}{kT} \overleftrightarrow{c_a{}^0 c_a} : \text{grad } \vec{u} + \left(\frac{m_a c_a^2}{2kT} - \frac{5}{2} \right) \text{grad } \ln T \cdot \vec{c_a} + \frac{n_a}{n} \vec{c_a} \cdot \vec{d_{ab}} = \right\} =$$

$$f_a^{(0)} \left\{ C(\varphi_a^{(1)}) + C(\varphi_a^{(1)} \varphi_b^{(0)}) \right\} \tag{C.3}$$

where

$$\vec{d_{ab}} = -\vec{d_{ba}} = \text{grad } \frac{n_a}{n} + \frac{n_a n_b (m_a - m_b)}{n\rho} \text{grad } \ln p$$

$$- \frac{\rho_a \rho_b}{p\rho} (\vec{F_a^e} - \vec{F_b^e}) - \frac{n_a n_b}{p\rho} (m_b e_a - m_a e_b) \vec{E} \tag{C.4}$$

Just as in the magnetic field free case all the theorems showing that Eq. (C.3) has no unique solution but an infinite number composed by a solution to the inhomogeneous part plus as arbitrary linear combination to the solution of the homogeneous part in these case being m_i, $m_i\vec{c_i}$, and $\frac{1}{2}m_i c_i^2$ where $(i = a, b)$, hold true. The latter solution is uniquely determined by the subsidiary conditions so one gets that

$$-\varphi_i^{(1)} = C_a{}^0 C_a \mathbb{B}_i : \text{grad } \vec{u} + \mathbb{A}_i \vec{c_i} \cdot \text{ grad } \ln T + \mathbb{D}_i \vec{d_{ij}} \cdot \vec{c_i} \tag{C.5}$$

where $\mathbb{A}_i$, $\mathbb{B}_i$ and $\mathbb{D}_i$ are scalar functions of c_i, n_i, T, etc. $\mathbb{A}_i$ and $\mathbb{D}_i$ are still subjected to the subsidiary condition that

$$\sum_i m_i \int f_i^{(0)} \left\{ \begin{array}{c} \mathbb{A}_i \\ \mathbb{D}_i \end{array} \right\} c_i^2 d\vec{c_i} = 0 \tag{C.6}$$

These functions depend on the particular interaction potential between the ions and the electrons in the gas and are solutions to the equations

$$-\left(\frac{m_i c_i^2}{2kT} - \frac{5}{2}\right) f_u^{(0)} \vec{c_i} = \{C(\mathbb{A}_i \vec{c_i}) + C(\mathbb{A}_i \vec{c_i} + \mathbb{A}_j \vec{c_j})\} f_i^{(0)} \tag{C.6a}$$

and

$$-\frac{n_i}{n} \vec{c_i} f_i^{(0)} = f_i^{(0)} \{C(\mathbb{D}_i \vec{c_i}) + C(\mathbb{D}_i \vec{c_i} + \mathbb{D}_i \vec{c_i} + \mathbb{D}_j \vec{c_j})\} \tag{C.6b}$$

The solution for $\mathbb{B}_i$ associated with the tensor grad $\vec{u}$ will not couple with the vectorial fluxes $\vec{d_{ab}}$ and grad T so we shall simply ignore it in what follows. For isotropic systems there are no visco-electric effects. The minus sign in Eq. (C.5) has been included for convenience. Ignoring $\mathbb{B}_i$,

$$\varphi_i^{(1)} = -\mathbb{A}_i \vec{c_i} \cdot \frac{1}{T} \text{grad } T - \mathbb{D}_i \cdot \vec{c_i} \vec{d_{ij}} \tag{C.5'}$$

can now allow us to compute the different fluxes in the mixture. This we shall do in detail to clearly exhibit how does Onsager's reciprocity theorem holds true. There are essentially three currents, the heat flow, the mass flow and the electrical flow, the forces being grad T, $\vec{d}_{ij}^{(0)}$ and $\vec{d}_{ij}^{(e)}$ where

$$\vec{d}_{ab}^{(0)} = \text{grad } \frac{n_a}{n} + \frac{n_a n_b}{n\rho}(m_a - m_b) \text{ grad } \ln p \tag{C.7a}$$

$$\vec{d}_{ab}^{(e)} = \frac{n_a n_b}{p\rho}(m_b e_a - m_a e_b)\vec{E} \tag{C.7b}$$

for simplicity other external forces, $\vec{F}_i^{(e)} = 0$. Therefore, $\vec{d}_{ij}^{(0)}$ must be related to conventional diffusion processes and $\vec{d}_{ij}^{(e)}$ with electrical phenomena together with their respective cross effects. For this purpose we shall follow closely the treatment contained in references [2] and [3]. Onsager's symmetry arises then in a very simple way. We recall that

$$\vec{J_i} = m_i \int \vec{c_i} f_i(c_i) d\vec{c_i} = m_i n_i \langle \vec{c_i} \rangle = \rho_i \langle \vec{c_i} \rangle$$

$$\sum_i \vec{J_i} = 0 \quad (\vec{J_a} = -\vec{J_b})$$

Also

$$\vec{J_c} = n_a e_a \langle \vec{c_a} \rangle + n_b e_b \langle \vec{c_b} \rangle = \frac{e_a}{m_a} \vec{J_a} + \frac{e_b}{m_b} \vec{J_b} \tag{C.8a}$$

We now use the fact that the results of ref. [1] and [2] for a binary mixture do not depend on the explicit form of the diffusive force $\vec{d}_{ab}$. In particular, as shown in Chap. 2,

$$\vec{J} = \frac{1}{m_a}\vec{J}_a + \frac{1}{m_b}\vec{J}_b = \frac{m_b - m_a}{m_a m_b}\vec{J}_a \tag{C.8b}$$

$$\vec{J'}_q = \vec{J}_q - \frac{5}{2}kT\vec{J} \tag{C.8c}$$

as in the inert mixture and in addition we have the electric current

$$\vec{J}_e = \frac{e_a}{m_a}\vec{J}_a + \frac{e_b}{m_b}\vec{J}_b = \frac{e_b m_a + e_a m_b}{m_a m_b}\vec{J}_a \tag{C.8d}$$

On the other hand, as shown in Ref. [4], one may write that,

$$\vec{J}_a = -L_{aq}\text{ grad } \ln T - L_{ab}\vec{d}_{ab} \tag{C.9}$$

and L_{aq}, L_{ab} obey the OR theorem. It is also clear that by (C.8b) and (C.8d) the expressions for $\vec{J}$ and $\vec{J}_e$ will also fulfill such theorem so indeed we have the three linear relations,

$$\begin{aligned} \vec{J} &= -L_{aa}\vec{d}^{(0)}_{ab} - L_{ae}\vec{d}^{(0)}_{ab} - L_{aq}\text{ grad } T \\ \vec{J'}_q &= -L_{qa}\vec{d}^{(0)}_{ab} - L_{qe}\vec{d}^{(0)}_{ab} - L_{qq}\text{ grad } T \\ \vec{J}_c &= -L_{ea}\vec{d}^{(0)}_{ab} - L_{eq}\text{ grad } T - L_{ee}\vec{d}^{(0)}_{ab} \end{aligned} \tag{C.10}$$

and by its own structure, $\vec{d}^{(0)}_{ab} \propto -\text{grad } \phi$ since $\vec{E} = -\text{grad } \phi$.

The reader may now see that we may write the Onsager matrix in a more canonical form, namely

$$\begin{pmatrix} \vec{J} \\ \vec{J}_c \\ \vec{J}_q \end{pmatrix} = \begin{pmatrix} L_{aa} & L_{ae} & L_{aq} \\ L_{ea} & L_{ee} & L_{eq} \\ L_{qa} & L_{qe} & L_{qq} \end{pmatrix} \begin{pmatrix} -\vec{d}^{(0)}_{ab} \\ -\vec{d}^{(e)}_{ab} \\ -\text{ grad } T \end{pmatrix} \tag{C.11}$$

Although Eqs. (C.10) are those which follow from microscopic reversibility and are therefore the holders of symmetry, in Eq. (C.11) it is no longer obvious that indeed one can make $L_{ae} = L_{ea}$, etc. but only their ratio can be shown to be constant. Nevertheless in their comparison with experiment the relations following from Eq. (C.11) as well as (C.8b-d) will be used. This point has to be kept in mind. (See Ref. [5] Chap. 9). We shall come back to this point later on.

We now proceed with the evaluation of $\vec{J}_a (= -\vec{J}_b)$ and $\vec{J}_q$ in order to evaluate the corresponding transport coefficients. For this purpose we shall assume that

$$\begin{aligned} \mathbb{A}_i(c_i, \ldots) &= \textstyle\sum_{p=0}^{\infty} a_i^{(p)} S_{\frac{3}{2}}^{(p)}(c_i) \\ \mathbb{D}_i(c_i, \ldots) &= \textstyle\sum_{q=0}^{\infty} d_i^{(p)} S_{\frac{3}{2}}^{(q)}(c_i) \end{aligned} \tag{C.12}$$

$S_{\frac{3}{2}}^{(m)}(c_i)$ are Sonine polynomials whose properties are summarized in p. 35 of the main text. First of all the subsidiary conditions (C.6) now read,

$$\sum_{p=0}^{\infty} \left\{ \begin{array}{c} a_i^{(p)} \\ d_i^{(p)} \end{array} \right\} \int f_i^{(0)} c_i^2 S_{\frac{3}{2}}^{(p)}(c_i) d\vec{c}_i = 0$$

Using the dimensionless velocity,

$$\vec{\omega}_i = \vec{c}_i \sqrt{\frac{m_i}{2kT}}$$

and

$$f_i^{(0)} = n_i \left(\frac{m_i}{2\pi kT} \right)^{3/2} \exp^{-\omega_i^2}$$

dropping irrelevant constants we get that

$$\sum_i n_i \left\{ \begin{array}{c} a_i^{(p)} \\ d_i^{(p)} \end{array} \right\} \delta_{p,0} = 0$$

which yields

$$\begin{aligned} n_a a_a^{(0)} + n_b a_b^{(0)} &= 0 \\ n_a d_a^{(0)} + n_b d_b^{(0)} &= 0 \end{aligned} \tag{C.13}$$

the resulting conditions to be imposed on the expansions in (C.12).

Using Eq. (C.5′) and (C.12) in the expression for $\vec{J}_a$ we find that

$$\vec{J}_a = -n_a kT \frac{\partial \ln T a_a^{(0)}}{\partial \vec{r}} - n n_a kT d_a^{(0)} \vec{d}_{ab} \tag{C.14a}$$

Introducing this last equation into Eq. (C.8a) and Eq. (C.8b) we can calculate $\vec{J}_c$ and the mass flow, namely,

$$\vec{J}_c = -n_a kT \left(\frac{m_a + m_b}{m_a m_b}\right) e a_a^{(0)} \frac{\partial \ln T}{\partial \vec{r}} - n n_a kT \left(\frac{m_a + m_b}{m_a m_b}\right) e d_a^{(0)} \vec{d}_{ab} \quad \text{(C.14b)}$$

and

$$\vec{J} = \frac{m_b - m_a}{m_a m_b} \left(-n_a kT a_a^{(0)} \frac{\partial \ln T}{\partial \vec{r}} - n n_a kT d_a^{(0)} \vec{d}_{ab}\right) \quad \text{(C.14c)}$$

Finally, using these two last equations and Eq. (C.13) in Eq. (C.8c) we have that,

$$\vec{J'}_q = \frac{5}{2}(kT)^2 \left(\frac{n_a a_a^{(1)}}{m_a} + \frac{n_b a_b^{(1)}}{m_b}\right) \text{ grad } \ln T - \left(\frac{m_a + m_b}{m_a m_b}\right) n n_a (kT)^2 d_a^{(0)} \vec{d}_{ab} \quad \text{(C.14d)}$$

Since in these equations $\vec{d}_{ab} = \vec{d}_{ab}^{(0)} + \vec{d}_{ab}^{(e)}$ and $\vec{E} = -\text{grad } \phi$ we have the transport coefficients matrix in which the diffusive force $\vec{d}_{ab}$ reduces to grad n_a if p is constant.

Insertion of the explicit form for $\vec{d}_{ab}$ in Eqs. (C.14b-d) leads precisely to the general set of equations (C.10). Notice should be made of the fact that Onsager's symmetry cannot be expected any longer. The proof of the ORT requires the full form of $\phi_i^{(1)}$ of which only four coefficients survive for the explicit calculation of the transport coefficients namely, $a_a^{(0)}$, $d_a^{(0)}$, $a_a^{(1)}$ and $a_b^{(1)}$. These will be obtained by solving the integral equations which arise when (C.5′) is inserted back in the linearized Boltzmann equation, Eq. (C.3).

Indeed, if this substitution is performed, we get that

$$\begin{aligned} -\left(\frac{m_a c_a^2}{2kT} - \frac{5}{2}\right) f_a^{(0)} \vec{c}_a &= \{C(\mathbb{A}_a \vec{c}_a) + C(\mathbb{A}_a \vec{c}_a + \mathbb{A}_b \vec{c}_b)\} f_a^{(0)} \\ -\frac{n_a}{n} f_a^{(0)} \vec{c}_a &= \{C(\mathbb{D}_a \vec{c}_a) + C(\mathbb{D}_a \vec{c}_a + \mathbb{D}_b \vec{c}_b)\} f_a^{(0)} \end{aligned} \quad \text{(C.15)}$$

and two identical equations for species b. These equations, far simpler than those arising in the $B \neq 0$ case will be solved by a rather straightforward variational method using the results of Appendix B. The argument goes as follows: Eqs. (C.15) are of the form,

$$R_i(\vec{v}) = \sum_{j=a}^{b} \int d\vec{v'}_i g\sigma(\Omega) d\Omega f_i^{(0)}(\vec{v}) f_j^{(0)}(\vec{v}_1) \left[T_i(\vec{v'}_i) + T_j(\vec{v'}_j) - T_i(\vec{v}_i) + T_j(\vec{v}_j)\right] \quad \text{(C.16)}$$

where $R_i(\vec{v})$ is known, T_i is unknown and $\vec{v'}_i$ and $\vec{v'}_j$ are the velocities of the particles after collision, and for collisions among of the same species $(\vec{v'}, \vec{v'}_1 \rightarrow \vec{v}, \vec{v}_1)$. Also,

$$C(\mathbb{A}_a\vec{c}_a) = \int d\vec{v'}_a g\sigma(\Omega)d\Omega \left\{\vec{c'}_a\mathbb{A}_a(\vec{c'}_a) + \vec{c'}_{a_1}\mathbb{A}_a(\vec{c'}_{a_1}) - \vec{c}_a\mathbb{A}_a(\vec{c}_a)\right. \\ \left. -\vec{c}_{a1}\mathbb{A}_a(\vec{c}_{a1})\right\} f_a^{(0)}(\vec{c}_a) f_a^{(0)}(\vec{c'}_a)$$

$$C(\mathbb{A}_a\vec{c}_a + \mathbb{A}_b\vec{c}_b) = \int d\vec{v'}_a g\sigma(\Omega)d\Omega \left\{\vec{c'}_a\mathbb{A}_a(\vec{c'}_a) + \vec{c'}_b\mathbb{A}_b(\vec{c'}_b) - \vec{c}_a\mathbb{A}_a(\vec{c}_a)\right. \\ \left. -\vec{c}_b\mathbb{A}_b(\vec{c}_b)\right\} f_a^{(0)}(\vec{c}_a) f_a^{(0)}(\vec{c}_b)$$

and in each case $g =\mid \vec{v}_i - \vec{v}_j \mid$, the relative velocity of the colliding particles.

Let now $t_i(\vec{v})$ be a trial function proposed as solution to Eq. (C.15). Integrating over $\vec{v}_i$ after multiplication by $t_i(\vec{v})$, one gets

$$\sum_i \int d\vec{v}_i R_i(\vec{v}) : t_i(\vec{v}) = \sum_{ij} \int d\vec{v}_i \vec{v}_j g\sigma(\Omega) d\Omega f_i^{(0)}(\vec{v}_i) f_j^{(0)}(\vec{v}_j) \\ t_i(\vec{v}) : \left[t_i(\vec{v'}_i) + t_j(\vec{v'}_j) - t_i(\vec{v}_i) + t_j(\vec{v}_j)\right]$$

This equation according to the definition of collisional brackets and Eq. (B.8) yield that

$$\sum_i \int d\vec{v}_i R_i(\vec{v}) : t_i(\vec{v}) = -\frac{1}{2}\{t_i, t_i\} =$$

$$-n_a^2 \left[t_a, t_a\right]_{aa} - n_a n_b \left[t_a + t_b, t_a + t_b\right]_{ab} - n_b^2 \left[t_b, t_b\right]_{bb} \qquad \text{(C.17)}$$

Clearly, from Eq. (C.16)

$$\sum_i \int d\vec{v}_i R_i(\vec{v}) : t_i(\vec{v}) = -\frac{1}{2}\{t_i, T_i\}$$

so if $t_i(\vec{v})$ in fact satisfies the integral equation (C.15) we would have that

$$\{t_i, t_i\} = \{t_i, T_i\}$$

By Eq. (B.7),

$$\{t_i - T_i; t_i - T_i\} = \{t_i, t_i\} - 2\{t_i, T_i\} + \{T_i, T_i\} \geq 0$$

then

$$-2\sum_i \int d\vec{v}_i R_i(\vec{v}) : t_i(\vec{v}) = \{t_i, t_i\} \leq \{T_i, T_i\} \qquad \text{(C.18)}$$

Eq. (C.18) is the basis of the variational principle. The proposed solution $t_i(\vec{v})$ must be such that it maximizes the collision integral $\{T_i, T_i\}$.

We start with the second of Eqs. (C.15) which after multiplication by $f_a^{(0)}$ on both sides and summation over the two species, reads

$$-2\sum_i \int d\vec{c}_i t_i : R_i = -\frac{2}{n}\sum_i n_i \int d\vec{c}_i \vec{c}_i \vec{c}_i f_i^{(0)}(\vec{c}_i) \sum_{p=0}^{\infty} d_p^{(i)} S_{\frac{3}{2}}^p(\omega_i)$$

after substitution of the trial function

$$\vec{t}_i = \vec{c}_i \sum_{p=0}^{\infty} d_i^{(p)} S_{\frac{3}{2}}^p(\vec{c}_i)$$

Introducing the dimensionless velocity $\vec{\omega}_i$ and carrying the integrals recalling that

$$\int_0^\infty d\omega \omega^4 \exp^{-\omega^4} S_{\frac{3}{2}}^p S_{\frac{3}{2}}^0 = \frac{3}{8}\sqrt{\pi}\delta_{p,0}$$

and introducing the subsidiary condition given by Eq. (C.13) we finally arrive at the result that

$$-2\sum_i \int d\vec{c}_i t_i : R_i = -6kT \frac{n_a}{n} \frac{(n_a m_b - n_b m_a)}{m_a m_b} d_a^{(0)} \qquad \text{(C.19)}$$

The inequality (C.18) thus yields that

$$-6kT \frac{n_a}{n} \frac{(n_a m_b - n_b m_a)}{m_a m_b} d_a^{(0)} = \left\{ \vec{c}_a \sum_p d_a^{(p)} S_{\frac{3}{2}}(\omega_a); \vec{c}_a \sum_p d_a^{(p)} S_{\frac{3}{2}}(\omega_a) \right\}$$

If we now use Eq. (B.8) for the binary mixture, to first approximation, noticing that $[\vec{c}_a, \vec{c}_a]_{aa} = [\vec{c}_b, \vec{c}_b]_{bb} = 0$ we have that ($d_a^{(p)} = 0$, $p > 0$) so that

$$-6kT \frac{n_a}{n} \left(\frac{n_a m_b - n_b m_a}{m_a m_b} \right) d_a^{(0)} = 2n_b \left[d_a^{(0)2} [\vec{\omega}_a, \vec{\omega}_a]_{aa} + 2d_a^{(0)} d_b^{(0)} [\vec{\omega}_a, \vec{\omega}_b]_{ab} + d_b^{(0)2} [\vec{\omega}_b, \vec{\omega}_b]_{bb} \right]$$

Once more, since $d_b^{(0)} = -(n_a/n_b) d_a^{(0)}$ we have

$$-3kT \frac{(n_a m_b - n_b m_a)}{n(m_a m_b)} d_a^{(0)} = n_b d_a^{(0)2} \left[[\vec{\omega}_a, \vec{\omega}_a]_{aa} - \frac{n_a}{n_b} [\vec{\omega}_a, \vec{\omega}_b]_{ab} + \frac{n_a^2}{n_b^2} [\vec{\omega}_b, \vec{\omega}_b]_{bb} \right]$$

so that the trivial solution is $d_a^{(0)} = 0$ and the sought one is given by

$$d_a^{(0)} = -\frac{3kT}{n}\left(\frac{n_a m_b - n_b m_a}{m_a m_b}\right)\left(1 - 2\frac{n_a}{n_b}\sqrt{\frac{m_a}{m_b}} + \frac{n_a^2}{n_b^2}\right)^{-1} \frac{1}{[\vec{\omega}_a, \vec{\omega}_b]_{ab}} \quad \text{(C.20a)}$$

As we see everything may be expressed in terms of a single collision integral using the results of Appendix D, namely

$$[\vec{\omega}_a, \vec{\omega}_b]_{ab} = -\sqrt{\frac{m_a}{m_b}}\,[\vec{\omega}_a, \vec{\omega}_a]_{ab} \quad \text{(C.20)}$$

and

$$[\vec{\omega}_b, \vec{\omega}_b]_{ab} = \frac{m_a}{m_b}\,[\vec{\omega}_a, \vec{\omega}_a]_{ab}$$

where

$$[\vec{\omega}_a, \vec{\omega}_a]_{ab} = \frac{1}{(4\pi\epsilon_0)^2}\frac{\sqrt{2\pi}e^4}{(kT)^{\frac{3}{2}}}\frac{1}{\sqrt{m_a}}\ln\left(1 + \left(\frac{4kTd}{e^2}\right)^2\right) \quad \text{(C.20b)}$$

and

$$d = \sqrt{\frac{kT\epsilon_0}{e^2 n}} \quad \text{(C.20c)}$$

is Debye's length. Eqs (C.20a-b-c) are the final result of this calculation. Improvement on the values for d_0^a can be obtained by considering more terms in the evaluation of $\{\quad\}$ but we leave that to the reader if and when he considers it necessary.

For the first of the equations (C.15) we have that

$$-2\sum_i \int d\vec{c}_i t_i : R_i = -2\sum_i \int \sum_{p=0}^{\infty} a_i^{(p)} n_i \frac{m_i}{2\pi kT}^{\frac{3}{2}} 4\pi \left(\frac{m_i c_i^2}{2kT} - \frac{5}{2}\right)$$

$$\exp^{-\frac{m_i c_i^2}{2kT}} c_i^4 dc_i S_{\frac{3}{2}}^{(p)}(\omega_i)$$

where the trial function used is the one in the previous case with different coefficients. Introducing the velocity ω_i, recalling that

$$\int_0^{\infty} S_{\frac{3}{2}}^{(1)}(\omega) S_{\frac{3}{2}}^{(p)}(\omega)\omega^4 \exp^{-\omega^2} d\omega = \frac{15}{16}\sqrt{\pi}\delta_{p,1}$$

we wind up with the result that

$$15kT\left(\frac{n_a}{m_a}a_a^{(1)} + \frac{n_b}{m_b}a_b^{(1)}\right) = \{t, t\}$$

The calculation of $\{t, t\}$ is a rather cumbersome procedure. One resorts to Eq. (B.3) setting

$$K_a = L_a = a_0^{(a)}\vec{c}_a + a_a^{(1)}S_{\frac{3}{2}}(\vec{c}_a)$$

$$K_b = L_b = a_0^{(b)}\vec{c}_b + a_b^{(1)}S_{\frac{3}{2}}(\vec{c}_b)$$

and makes use of all the values for the corresponding collision integrals which are given in Appendix D. After a lengthy but straight forward algebraic procedure one finds that $\{t_i, t_i\}\,\tau$ is given precisely by the terms arising from the collisional integrals which are written in the right hand side of Eq. (5.8) after the ω-dependent terms namely the second and third rows are ignored. We find unnecessary to repeat that equation here. The remaining steps of the variational procedure lead precisely to Eqs. (5.9) and (5.10) when $\omega_a = \omega_b = 0$. As expected the results in this case are those quoted in Eqs. (5.11). The procedure to obtain the d_i's is completely analogous.

Bibliography

[1] L. Spitzer; *The Physics of Fully Ionized Gases*, Wiley-Interscience, New York (1962).

[2] S. Chapman and T. G. Cowling; loc. cit. Chap. 1.

[3] G. W. Ford and G. E. Uhlenbeck; loc. cit. Chap. 1.

[4] P. Goldstein and L. García-Colín; *J. Non-Equilib. Thermodyn.* **30**, 173 (2005).

[5] S. R. de Groot and P. Mazur; *Non-Equilib. Thermodynamics*, Dover Publications Inc., Mineola, N.Y. (1984), Chap. XI.

[6] J. O. Hirschfelder, C. F. Curtiss and R. B. Byrd; *The Molecular Theory of Liquids and Gases*, John Wiley & Sons, New York (1964). 2nd ed.

Appendix D

The Collision Integrals

In order to determine the coefficients of viscosity, thermal conduction, and diffusion of a gas, it is necessary first to evaluate the collision integrals. In this Appendix we shall consider collisions and the evaluation of the various collision integrals in detail and how the cross section can be simply expressed for Coulomb interactions.

Suppose we have two particles colliding, the first of mass m_1, charge e_1, velocity v_1 and the second mass m_2, charge e_2 and velocity v_2. For Coulomb forces between the particles the equation of motion is given by

$$\mu \frac{d^2 \vec{r}}{dt^2} = \frac{k_e e_1 e_2}{r^3} \vec{r} \tag{D.1}$$

where

$$\mu = \frac{m_1 m_2}{m_1 + m_2}$$

is the reduced mass, and $k_e = 1/4\pi\varepsilon_0$ where ε_0 the permitivity of vacuum ($\varepsilon_0 = 8.85 \times 10^{-12}$ C^2/Jm).

The geometry of the collision is shown in Figure D.1. Initially particle 1 has a relative velocity

$$\vec{g} = \vec{v}_1 - \vec{v}_2 \tag{D.2}$$

and asymptotic distance of approach b. We suppose that its relative position makes an angle β with the direction $-\vec{g}$, so initially $\beta = 0$. Finally $\beta = \pi - \chi$ where χ is the scattering angle we wish to find as a function of g and b.

The differential cross section for scattering into unit solid angle shall be

L.S. García-Colín, L. Dagdug, *The Kinetic Theory of a Dilute Ionized Plasma*, Springer Series on Atomic, Optical, and Plasma Physics 53

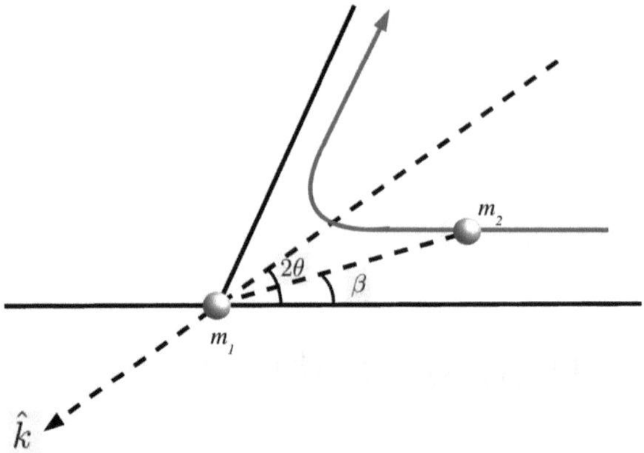

Figure D.1: Geometry of a collision.

replaced by the well known Rutherford formula, namely,

$$\sigma(\chi, \varepsilon) = \left(\frac{k_e e_1 e_2}{2\mu g^2} \operatorname{cosec}^4 \frac{1}{2}\chi \right)^2 \tag{D.3}$$

The angles of this expression are depicted in Figure D.1.

From Figure D.1 we see that,

$$\begin{aligned} \chi &= \pi - 2\theta \\ \cot \tfrac{1}{2}\chi &= \frac{\mu b g^2}{k_e e_1 e_2} \end{aligned} \tag{D.4}$$

is convenient to make use of the unit vector $\hat{k}$ drawn as show in Figure D.1. Clearly

$$\hat{k} \cdot \vec{g} = \vec{g} \cos\theta = -\hat{k} \cdot \vec{g}\,' \tag{D.5}$$

where

$$\vec{g}\,' = \vec{v}\,'_2 - \vec{v}\,'_1 = \vec{c}\,'_2 - \vec{c}\,'_1 \tag{D.6}$$

and it is the relative velocity after the collision. Also

$$\begin{aligned} \vec{g}\,' &= \vec{g} - 2(\vec{g} \cdot \hat{k}) \\ &= \vec{g} - 2g\cos\theta \hat{k} \end{aligned} \tag{D.7}$$

It is also convenient to introduce the dimensionless numbers

$$M_1 = \frac{m_1}{m_1 + m_2}$$
$$M_2 = \frac{m_2}{m_1 + m_2} \tag{D.8}$$

The center of gravity velocity is, relative to the drift velocity $\vec{u}$

$$\vec{G} = M_1\vec{c}_1 + M_2\vec{c}_2 \tag{D.9}$$

and from (D.8) and (D.9) the following equations can be derived

$$\vec{c}_1 = \vec{G} + M_2\vec{g}$$
$$\vec{c}_2 = \vec{G} + M_1\vec{g}$$
$$\vec{c}'_1 = \vec{G} + M_2\vec{g}'$$
$$\vec{c}'_2 = \vec{G} + M_1\vec{g}' \tag{D.10}$$

Also,

$$\vec{c}'_1 = \vec{c}_1 + 2\vec{g}M_2 \cos\theta\hat{k}$$
$$\vec{c}'_2 = \vec{c}_2 + 2\vec{g}M_1 \cos\theta\hat{k} \tag{D.11}$$

Now define new variables $\vec{x}$ and $\vec{y}$ by

$$\vec{x} = \vec{g}\sqrt{\frac{\mu}{2kT}}$$
$$\vec{y} = \vec{G}\sqrt{\frac{m_1 + m_2}{2kT}} \tag{D.12}$$

Then

$$\vec{w}_1 = M_1^{\frac{1}{2}}\vec{y} + M_2^{\frac{1}{2}}\vec{x}$$
$$\vec{w}_2 = M_2^{\frac{1}{2}}\vec{y} - M_1^{\frac{1}{2}}\vec{x}$$
$$\vec{w}\,'_1 = w_1 - 2M_2^{\frac{1}{2}}\vec{x}\cos\theta\hat{k} \tag{D.13}$$
$$\vec{w}\,'_2 = w_2 - 2M_1^{\frac{1}{2}}\vec{x}\cos\theta\hat{k}$$

where

$$\vec{w}_i = \sqrt{\frac{m_i}{2kT}} \tag{D.14}$$

Since the Jacobian of the transformation is,

$$J = \frac{\partial(\vec{x},\vec{y})}{\partial(\vec{c}_1,\vec{c}_2)} = \frac{\sqrt{m_1 m_1}}{2kT} \tag{D.15}$$

and from Eq. (D.4) in terms of this new variables, the collision integral can be written as,

$$[G_1, H_2; K_1 L_2]_{12} = -\left(\frac{2kT}{\mu}\right)^{\frac{1}{2}} \frac{1}{\pi^3} \int d\vec{x} d\vec{y} d\varepsilon b db \vec{x} \exp(-x^2 - y^2)$$
$$[G_1(\vec{w}_1) + H_2(\vec{w}_2)] : [K_1(\vec{w}\,'_1) + L_2(\vec{w}\,'_2) - K_1(\vec{w}_1) + L_2(\vec{w}_2)] \tag{D.16}$$

Using this formula all the integrals for collisions between an electron and an ion can be worked out. Then the integrals for collisions between like particles can be obtained by setting the masses equal. We shall give the details of just one calculation, namely of $[\vec{w}_1, \vec{w}_1]_{12}$. From Eq. (D.16) this is

$$[\vec{w}_1, \vec{w}_1]_{12} = -\left(\frac{2kT}{\mu}\right)^{\frac{1}{2}} \frac{1}{\pi^3} \int \cdots \int d\vec{x} d\vec{y} d\varepsilon b db \vec{x} \exp(-x^2 - y^2) \vec{w}_1 \cdot (\vec{w}\,'_1 - \vec{w}_1) \tag{D.17}$$

Now from Eq. (D.11)

$$\vec{w}_1 \cdot (\vec{w}\,'_1 - \vec{w}_1) = -\left[M_1^{\frac{1}{2}}\vec{y} + M_2^{\frac{1}{2}}\vec{x}\right] 2M_2^{\frac{1}{2}}\vec{x}\cos\theta \cdot \hat{k} \tag{D.18}$$

Since the y integral is odd, it averages to zero on integrating over $\vec{y}$, and finally

$$[\vec{w}_1, \vec{w}_1]_{12} = 2M_2 \left(\frac{2kT}{\mu}\right)^{\frac{1}{2}} \frac{2\pi}{\pi^{\frac{3}{2}}} \int d\vec{x} x^3 \exp(-x^2) \int b db \cos^2\theta \tag{D.19}$$

Using Eq. (D.4) this last integral is

$$\int_0^d bdb\cos^2\theta = \int_0^d \frac{bdb}{1+\left(\dfrac{2kTx^2}{k_e e_1 e_2}\right)^2 b^2}$$
$$= \frac{1}{2}\left(\frac{k_e e_1 e_2}{2kT}\right)^2 \frac{1}{x^4}\ln\left\{1+d^2\left(\frac{2kTx^2}{k_e e_1 e_2}\right)^2\right\} \tag{D.20}$$

This expression diverges if d goes to infinity but the divergence is only logarithmic and is therefore very slow. Whence, it does not matter very much what choice we make for d within reasonable limits. Because the value we should use for d is not fixed precisely and because the answer is insensitive anyway we might just as well replace x^2 where it appears inside the logarithm by its average value which is 2. Hence Eq. (D.20) becomes

$$\frac{1}{2}\left(\frac{k_e e_1 e_2}{2kT}\right)^2 \frac{1}{x^4}\psi$$

where ψ is the logarithm factor

$$\psi = \ln\left\{1+\left(\frac{4kTd}{k_e e_1 e_2}\right)^2\right\} \tag{D.21}$$

So Eq. (D.19) becomes

$$[\vec{w}_1,\vec{w}_1]_{12} = M_2\left(\frac{2kT}{\mu}\right)^{\frac{1}{2}}\left(\frac{k_e e_1 e_2}{2kT}\right)^2 \frac{2\psi}{\sqrt{\pi}}\int_0^\infty d\vec{x}\frac{1}{x}\exp(-x^2)$$
$$= \sqrt{2\pi}\frac{(k_e e_1 e_2)^2\psi}{(kT)^{\frac{3}{2}}}\left\{\frac{m_2}{m_1(m_1+m_2)}\right\}^{\frac{1}{2}} \tag{D.22}$$

If m_2 it is much grater than m_1

$$[\vec{w}_1,\vec{w}_1]_{12} \simeq \sqrt{2\pi}\frac{(k_e e_1 e_2)^2\psi}{\sqrt{m_1}(kT)^{\frac{3}{2}}} = \varphi \tag{D.23}$$

Interchanging the masses in Eq. (D.23) gives

$$[\vec{w}_2,\vec{w}_2]_{12} \simeq \sqrt{2\pi}\frac{(k_e e_1 e_2)^2\psi}{(kT)^{\frac{3}{2}}}\left\{\frac{m_1}{m_2(m_1+m_2)}\right\}^{\frac{1}{2}} = \frac{m_1}{m_1}[\vec{w}_1,\vec{w}_1]_{12} \tag{D.24}$$

Now we shall calculate $[\vec{w}_1, \vec{w}_2]_{12}$ as follows. By definition

$$[\vec{w}_1, \vec{w}_2]_{12} = -\left(\frac{2kT}{\mu}\right)^{\frac{1}{2}} \frac{1}{\pi^3} \int d\vec{x}d\vec{y}d\varepsilon bdb\vec{x}\exp(-x^2 - y^2)\vec{w}_1 \cdot (\vec{w}\,'_2 - \vec{w}_2) \tag{D.25}$$

and from Eq. (D.13)

$$\vec{w}\,'_2 - \vec{w}_2 = -\left(\frac{m_1}{m_2}\right)^{\frac{1}{2}} (\vec{w}\,'_1 - \vec{w}_1) \tag{D.26}$$

Comparing Eqs. (D.25) and (D.26) with Eq. (D.22) we therefore see that

$$[\vec{w}_1, \vec{w}_2]_{12} = -\left(\frac{m_1}{m_2}\right)^{\frac{1}{2}} [\vec{w}_1, \vec{w}_1]_{12} \tag{D.27}$$

From Eq. (D.22) and Eq. (D.27)

$$[\vec{w}_1, 0; \vec{w}_1, \vec{w}_2]_{12} = \left\{1 - \left(\frac{m_1}{m_2}\right)^{\frac{1}{2}}\right\} \sqrt{2\pi}\frac{(k_e e_1 e_2)^2\psi}{(kT)^{\frac{3}{2}}} \left\{\frac{m_2}{m_1(m_1 + m_2)}\right\}^{\frac{1}{2}} \tag{D.28}$$

We can now get an expression for $[\vec{w}_1, \vec{w}_1]_1$ by setting $m_1 = m_2$ in Eq. (D.28). This gives

$$[\vec{w}_1, \vec{w}_1]_{11} = 0 \tag{D.29}$$

Similarly

$$[\vec{w}_2, \vec{w}_2]_{22} = 0 \tag{D.30}$$

In a similar way all other integrals we need can be evaluated and the results are given in the following list.

$$[\vec{w}_1, \vec{w}_1]_{11} = 0$$

$$[\vec{w}_2, \vec{w}_2]_{22} = 0$$

$$[\vec{w}_1, \vec{w}_1]_{12} = \varphi$$

$$[\vec{w}_1, \vec{w}_2]_{12} = -M_1^{\frac{1}{2}}\varphi$$

$$[\vec{w}_2, \vec{w}_2]_{12} = M_1\varphi$$

$$[\vec{w}_1, \vec{w}_1 S^1_{\frac{3}{2}}(\vec{w}_1^2)]_{12} = \frac{3}{2}\varphi$$

$$[\vec{w}_2, \vec{w}_2 S^1_{\frac{3}{2}}(\vec{w}_2^2)]_{12} = \frac{3}{2} M_1^2 \varphi$$

$$[\vec{w}_1, \vec{w}_2 S^1_{\frac{3}{2}}(\vec{w}_2^2)]_{12} = -\frac{3}{2} M_1^{\frac{3}{2}} \varphi$$

$$[\vec{w}_1 S^1_{\frac{3}{2}}(\vec{w}_1^2), \vec{w}_2]_{12} = -\frac{3}{2} M_1^{\frac{1}{2}} \varphi$$

$$[\vec{w}_2, \vec{w}_1 S^1_{\frac{3}{2}}(\vec{w}_1^2)]_1 = 0$$

$$[\vec{w}_2, \vec{w}_2 S^1_{\frac{3}{2}}(\vec{w}_2^2)]_2 = 0$$

$$[\vec{w}_1 S^1_{\frac{3}{2}}(\vec{w}_1^2), \vec{w}_1 S^1_{\frac{3}{2}}(\vec{w}_1^2)]_{12} = \frac{13}{4} \varphi$$

$$[\vec{w}_2 S^1_{\frac{3}{2}}(\vec{w}_2^2), \vec{w}_2 S^1_{\frac{3}{2}}(\vec{w}_2^2)]_{12} = \frac{15}{2} M_1^{\frac{1}{2}} \varphi$$

$$[\vec{w}_1 S^1_{\frac{3}{2}}(\vec{w}_1^2), \vec{w}_2 S^1_{\frac{3}{2}}(\vec{w}_2^2)]_{12} = -\frac{27}{4} M_1^{\frac{3}{2}} \varphi \left(1 + \frac{16}{27}\delta\right)$$

$$[\vec{w}_1 S^1_{\frac{3}{2}}(\vec{w}_1^2), \vec{w}_1 S^1_{\frac{3}{2}}(\vec{w}_1^2)]_1 = \sqrt{2}\varphi(1 - \delta)$$

$$[\vec{w}_2 S^1_{\frac{3}{2}}(\vec{w}_2^2), \vec{w}_2 S^1_{\frac{3}{2}}(\vec{w}_2^2)]_2 = \sqrt{2M_1}\varphi(1 - \delta)$$

where

$$\delta = \frac{1}{\psi} \frac{\left(\dfrac{4kTd}{k_e e_1 e_2}\right)^2}{1 + \left(\dfrac{4kTd}{k_e e_1 e_2}\right)^2}$$

which, for all except very extreme conditions is $1/\psi$ and very small compared to unity. Hence under almost all conditions δ can be set equal to zero in this list.

It remains to examine the validity of Eq. (D.1) which assumed that while the particles were interacting all forces other than their Coulomb interaction could be ignored. This will be valid provided the Debye distance, d, is smaller than the gyromagnetic radius, namely, provided

$$d = \left(\frac{kT}{4\pi(k_e e_1 e_2)}\right)^{\frac{1}{2}} << \frac{mv}{eB}$$

This inequality holds very well except for low densities $\leq 10^{12}$ m^{-3} and high fields ($H \geq 5000$ gauss).

Appendix E

Calculation of the Coefficients $a_i^{(0)}$, $a_i^{(1)}$, $d_i^{(0)}$ and $d_i^{(1)}$

We here outline the details of two calculations leading to the coefficients $a_i^{(1)} = a_i^{(1)(1)} + iBa_i^{(1)(2)}$, $i = a, b$; the same for $a_i^{(0)}$ and their analogs for $d_i^{(0)}$, $d_i^{(1)}$. The former ones are simply the solution to the inhomogeneous system of three equations defined in Eq. (5.9) in the text. This system can be solved by hand with some approximations which are thereafter compared with the ones obtained with a computer program. After separating their real and complex parts one obtains that

$$\begin{aligned}
Re\left(a_a^{(0)}\right) &= a_a^{(1)(0)} = && \tfrac{5\tau}{\Delta_1}\left(1.956 - 19.8x^2 - 4.5x^4\right) \\
Im\left(a_a^{(0)}\right) &= a_a^{(2)(0)} = && \tfrac{5\tau}{\Delta_1}\left(21.3x + 4.78x^3\right) \\
Re\left(a_a^{(1)}\right) &= a_a^{(1)(1)} = && \tfrac{5\tau}{\Delta_1}\left(1.305 + 68.28x^2 + 16.14x^4\right) \\
Im\left(a_a^{(1)}\right) &= a_a^{(1)(2)} = && \tfrac{5\tau}{\Delta_1}\left(5.31x + 82.452x^3 + 18x^5\right) \\
Re\left(a_b^{(1)}\right) &= a_b^{(1)(1)} = && \tfrac{1}{M_1}\tfrac{5\tau}{\Delta_1}\left(0.35154 + 7.21x^2 + 37.99x^4\right) \\
Im\left(a_b^{(1)}\right) &= a_b^{(1)(2)} = && \tfrac{1}{M_1}\tfrac{5\tau}{\Delta_1}\left(0.036x - 16.45x^3 - 18x^5\right)
\end{aligned} \tag{E.1}$$

where

$$\Delta_1 = 3.744 + \times 10^{10}x^2 + 4.59 \times 10^{10}x^4 + 90x^6 \tag{E.2}$$

As stated in page 49 the only difference between the previous case and the one corresponding to the $\mathbb{D}_i$ functions in Eq. (3.15b) is that in constructing

L.S. García-Colín, L. Dagdug, *The Kinetic Theory of a Dilute Ionized Plasma*, Springer Series on Atomic, Optical, and Plasma Physics 53

the variational procedure the inhomogeneous terms in Eqs. (5.10) change to the values given in Eqs. (5.9). Thus the algebraic system of inhomogeneous equations may be solved to yield the following results:

$$\begin{aligned}
Re\left(d_a^{(0)}\right) &= d_a^{(1)(0)} = & -\tfrac{3\tau}{2\Delta_2}\left(2.86 + 6.616x^2 + 1.415x^4\right) \\
Im\left(d_a^{(0)}\right) &= d_a^{(2)(0)} = & -\tfrac{3\tau}{2\Delta_2}\left(27.745x + 40.05x^3 + 7.5x^5\right) \\
Re\left(d_a^{(1)}\right) &= d_a^{(1)(1)} = & \tfrac{3\tau}{2\Delta_2}\left(0.6523 - 6.604x^2 - 1.5x^4\right) \\
Im\left(d_a^{(1)}\right) &= d_a^{(1)(2)} = & -\tfrac{3\tau}{2\Delta_2}\left(7.095x + 1.598x^3\right) \\
Re\left(d_b^{(1)}\right) &= d_b^{(1)(1)} = & \tfrac{3\tau}{2\Delta_2}\left(0.5426 + 0.339x^2 + 1.5x^4\right) \\
Im\left(d_b^{(1)}\right) &= d_b^{(1)(2)} = & \tfrac{3\tau}{2\Delta_2}\left(5.343x + 4.265x^3\right)
\end{aligned} \tag{E.3}$$

where $\Delta_2 = \Delta_1$.

Appendix F

The proof of the equalities given in Eq. (8.7) is just a matter of a careful lengthy manipulation. All follow the same procedure. For instance,

$$2\overleftrightarrow{\tau}_i^{(5)} = 2\overleftrightarrow{Q}_i^{(7)}$$

Using the form for the right hand and shown in Eq. (8.6) we take, say component xy. Next we sum over repeated indices γ, φ and λ. Take γ first and then φ keeping only the non zero terms according to the structure of ϵ_{ijk}. In the final summation over λ, the resulting expression is identical to the one obtained for the left hand side according to Eq. (8.2). To show an easier example, this same procedure gives that

$$2(Q_i^{(2)})_{xy} = B_x C_x C_z - B_y C_y C_z - B_z(C_x^2 - C_y^2)$$

which is identical to $2(\tau_i^{(2)})_{xy}$. Thus

$$(Q_i^{(2)})_{\alpha\beta} = (\tau_i^{(2)})_{\alpha\beta} \quad \text{for } \alpha \neq \beta$$

and for $\alpha = \beta$ the result follows at once.

The two middle expressions in Eq. (8.7) are also straight forward. Since $(Q_i^{(4)})_{\alpha\beta} = 0$ if $\alpha = \beta$ one readily sees that after expansion

$$(\overleftrightarrow{\tau}_i^{(3)})_{xy} = (\overleftrightarrow{Q}_i^{(3)})_{xy} - \frac{1}{3}B^2(\overleftrightarrow{Q}_i^{(1)})_{xy} - \frac{1}{3}(\overleftrightarrow{Q}_i^{(5)})_{xy}$$

and for diagonal components,

$$(Q_i^{(4)})_{\alpha\alpha} = (\vec{C} \cdot \vec{B})^2 - B^2 C^2$$

accounts for the extra term. The same holds for $(\overleftrightarrow{\tau}_i^{(4)})$.

L.S. García-Colín, L. Dagdug, *The Kinetic Theory of a Dilute Ionized Plasma*,
Springer Series on Atomic, Optical, and Plasma Physics 53

Appendix G

List of useful integrals

$$\int_0^{2\pi} \sin^2 \theta d\theta = \int_0^{2\pi} \cos^2 \theta d\theta = \pi$$

$$\int_0^{\pi} \sin \theta \cos^2 \theta d\theta = \frac{2}{3}$$

$$\int_0^{\pi} \sin^3 \theta \cos^2 \theta d\theta = \frac{4}{15}$$

$$\int_0^{2\pi} \sin^4 \theta \cos^4 \theta d\theta = \frac{3}{64}\pi$$

$$\int_0^{\pi} \sin^5 \theta d\theta - \frac{16}{15}$$

$$\int_0^{2\pi} \sin^2 \theta \cos^2 d\theta = \frac{\pi}{4}$$

$$\int_0^{\pi} \sin^4 \theta d\theta = \int_0^{\pi} \cos^4 \theta d\theta = \frac{3}{8}\pi$$

$$\int_0^{\pi} \sin^3 \theta d\theta = \frac{4}{3}$$

$$\int_0^{2\pi} \sin^2 \theta \cos^4 \theta d\theta = \frac{\pi}{8}$$

$$\int_0^{\pi} \sin \theta \cos^6 \theta d\theta = \frac{2}{7}$$

L.S. García-Colín, L. Dagdug, *The Kinetic Theory of a Dilute Ionized Plasma*,
Springer Series on Atomic, Optical, and Plasma Physics 53

Appendix H

The collision integrals used to arrive at Eq. (8.33) in the text are,

$$\left[\overleftrightarrow{w_a {}^0 w_a}, \overleftrightarrow{w_a {}^0 w_a}\right]_{aa} = \sqrt{2}\varphi$$

$$\left[\overleftrightarrow{w_a {}^0 w_a}, \overleftrightarrow{w_a {}^0 w_a}\right]_{ab} = 2\varphi$$

$$\left[\overleftrightarrow{w_b {}^0 w_b}, \overleftrightarrow{w_b {}^0 w_b}\right]_{bb} = \sqrt{2}M_1\varphi$$

$$\left[\overleftrightarrow{w_b {}^0 w_b}, \overleftrightarrow{w_b {}^0 w_b}\right]_{ab} = \frac{10}{3}M_1\varphi$$

$$\left[\overleftrightarrow{w_a {}^0 w_a}, \overleftrightarrow{w_b {}^0 w_b}\right]_{ab} = -\frac{4}{3}M_1\varphi$$

where

$$M_1 = \frac{m_a}{m_a + m_b}$$

Now we start from Eq. (8.33) in the text. Calculating $\delta\mathfrak{D}(\mathfrak{I}_i) = 0$, collecting terms in $\delta g_a^{(0)}$ and $\delta g_b^{(0)}$, performing obvious arithmetical simplifications, introducing the Larmor frequencies ω_i, we arrive at two equations,

$$-\frac{10\tau m_a}{4kT} = (40 i\omega_a\tau + 9.66)\, g_a^{(0)} + \frac{8}{3}M_1^2 g_b^{(0)}$$

$$-\frac{10\tau m_b}{kTM_1} = (-40 i\omega_a\tau + 243.98)\, g_b^{(0)} - \frac{8}{3}M_1^2 g_a^{(0)}$$

Noticing that $\omega_a = M_1\omega_b$ if $m_b >> m_a$ and neglecting obvious small terms we find that the determinant for this system of equations is, setting $x = \omega_a\tau$

$$\Delta = \left(47 + 1.6x^2 + 9.842x\right) \times 10^3$$

L.S. García-Colín, L. Dagdug, *The Kinetic Theory of a Dilute Ionized Plasma*,
Springer Series on Atomic, Optical, and Plasma Physics 53

Trivially then

$$g_a^{(0)} = \frac{10\tau m_a}{kT}\frac{258.4 - 40ix}{\Delta} \tag{H.1}$$

$$g_b^{(0)} = \frac{10\tau m_b}{M_1 kT}\frac{9.66 + 40ix}{\Delta} \tag{H.2}$$

When G_i is substituted by $P_i/2$ and B by $2B$ we obtain the solution to Eq. (8.20),

$$p_a^{(0)} = \frac{20m_a\tau}{kT}\frac{258.4 - 40ix'}{\Delta'} \tag{H.3}$$

$$p_b^{(0)} = \frac{20\tau m_a}{M_1 kT}\frac{9.66 + 40ix'}{\Delta'}m_b \tag{H.4}$$

where $\omega_a' = 2\omega_a$ and $\Delta' = \Delta$ when $\omega_a = 2\omega_a$. Finally, the solution to Eqs. (8.15a) is obtained when $B = 0$, $\omega_a = \omega_b = 0$

$$\begin{aligned} l_a^{(0)} &= 1.046\tfrac{m_a\tau}{kT} \\ l_b^{(0)} &= 0.039\tfrac{m_b\tau}{M_1 kT} \end{aligned} \tag{H.5}$$

and τ, the mean free collision time is $\tau = 1/\varphi n$.

Omitting the unnecessary superscript naught, the relation between these results and the Γ_i's is now summarized as follows,

$$\begin{aligned} l_i &= \Gamma_i^1 + B^2\Gamma_i^3 \\ Rep_i &= 2\Gamma_i^1 + B^2\Gamma_i^4 \\ Imp_i &= B(\Gamma_i^2 + B^2\Gamma_i^5) \\ Reg_i &= \Gamma_i^1 + B^2\Gamma_i^3 \\ Img_i &= \Gamma_i^2 \end{aligned} \tag{H.6}$$

so that $B^2\Gamma_i^5 = Imp_i - Img_i$, whereas the shear viscosity η is determined by l_i when $B = 0$, in this case all other coefficients vanish. This completes the solution to the problem.

Appendix I

List of Marshall's Equations and Notation

I.1 Equations

$$-\vec{d}_2 = \vec{d}_1 = \nabla(\frac{n_1}{n}) + \frac{n_1 n_2 (m_2 - m_1)}{pn\rho}\nabla p - \frac{\rho_1 \rho_2}{p\rho}(X_1 - X_2) -$$
$$\frac{n_1 n_2}{p\rho}(e_1 m_2 - e_2 m_1)\vec{E}' \qquad \text{(M3.14)}$$

$$\vec{\jmath} = -\{\nabla \log T\}\frac{1}{2}(2kT)^{\frac{1}{2}}\sum_i \frac{n_i e_i}{\sqrt{m_i}} a_i^{I,0} - \{\vec{H} \times \nabla \log T\}\frac{1}{2}(2kT)^{\frac{1}{2}}\sum_i \frac{n_i e_i}{\sqrt{m_i}} a_i^{II,0}$$
$$-\vec{H}\{\vec{H} \cdot \nabla \log T\}\frac{1}{2}(2kT)^{\frac{1}{2}}\sum_i \frac{n_i e_i}{\sqrt{m_i}} a_i^{III,0} - \{n\vec{d}_i\}\frac{1}{2}(2kT)^{\frac{1}{2}}\sum_i \frac{n_i e_i}{\sqrt{m_i}} e_i^{I,0}$$
$$\{\vec{H} \times n\vec{d}_i\}\frac{1}{2}(2kT)^{\frac{1}{2}}\sum_i \frac{n_i e_i}{\sqrt{m_i}} e_i^{II,0} - \{\vec{H}(\vec{H} \cdot n\vec{d}_1)\}\frac{1}{2}(2kT)^{\frac{1}{2}}\sum_i \frac{n_i e_i}{\sqrt{m_i}} e_i^{III,0}$$
$$\text{(M3.61)}$$

$$\vec{\jmath} = \sigma^I \vec{D}'' + \sigma^{II}\vec{D}^{\perp} + \sigma^{III}\vec{h} \times \vec{D}^{\perp} + \varphi^I\{\nabla T\}'' + \varphi^{II}\{\nabla T\}^{\perp} + \varphi^{III}\vec{h} \times \{\nabla T\}^{\perp}$$
$$\text{(M3.62)}$$

$$D = \vec{E} + \frac{1}{c}\vec{u} \times \vec{H} - \frac{m_2 - m_1}{n(e_1 m_2 - e_2 m_1)}\nabla p - \frac{m_1 m_2}{e(m_1 + m_2)}(X_1 - X_2)$$
$$-\frac{p\rho}{n_1 n_2 (e_1 m_2 - e_2 m_1)}\nabla(\frac{n_1}{n}) \qquad \text{(M3.63)}$$

L.S. García-Colín, L. Dagdug, *The Kinetic Theory of a Dilute Ionized Plasma*,
Springer Series on Atomic, Optical, and Plasma Physics 53

$$\sigma^I = \frac{ne^2\tau}{2m_1}1.931$$

$$\sigma^{II} = \frac{ne^2\tau}{2m_1}\frac{\omega^2\tau^2 + 1.802}{\omega^4\tau^4 + 6.282\omega^2\tau^2 + 0.933} \tag{M7.8}$$

$$\sigma^{III} = \frac{ne^2\tau}{2m_1}\frac{-\omega\tau(\omega^2\tau^2 + 4.382)}{\omega^4\tau^4 + 6.282\omega^2\tau^2 + 0.933}$$

where

$$\omega = -\frac{eH}{cm_1}$$

and

$$\tau = \frac{3}{\sqrt{2\pi}}\frac{\sqrt{m_1}(kT)^{\frac{3}{2}}}{ne^4\psi} \tag{M7.10}$$

$$\vec{q} = -\theta^I\{\nabla T\}'' + \theta^{II}\{\nabla T\}^{\perp} + \theta^{III}\vec{h}\times\{\nabla T\}^{\perp} + \xi^I\vec{D}'' + \xi^{II}\vec{D}^{\perp} + \xi^{III}\vec{h}\times\vec{D}^{\perp} \tag{M7.16}$$

$$\theta^I = \frac{n\tau k^2T}{m_1}3.59$$

$$\theta^{II} = \frac{n\tau k^2T}{m_1}\left\{\frac{0.458\omega^2\tau^2+3.01}{\omega^4\tau^4+6.282\omega^2\tau^2+0.933} + \frac{4.458}{\omega^2\tau^2+12.716}\right\} \tag{M7.19}$$

$$\theta^{III} = \frac{n\tau k^2T}{m_1}1.25\left\{\frac{\omega\tau}{\omega^2\tau^2+12.716} - \frac{\omega\tau(\omega^2\tau^2+6.2)}{\omega^4\tau^4+6.282\omega^2\tau^2+0.933}\right\}$$

I.2 Notation

$\vec{E}' = \vec{E} + \frac{1}{c}\vec{u}\times\vec{H}$

c : Velocity of light.

e_i^m : coefficients of the expansions.

$\vec{H}$: The magnetic field.

$\vec{h}$: A unit vector in the direction of $\vec{H}$.

i, j : Subscripts labeling electrons and ions, 1 for electrons and 2 for ions.

$\vec{j}$: The conduction current.

k : Boltzmann's constant.

m_i : The mass of particle i.

n_i : The number density of particles i.

$n = n_1 + n_2$: The total number density.

$\vec{q}$: The heat flux vector.

T : Temperature.

$\vec{u}$: The drift velocity.

X_i : any non-electromagnetic force per unit mass which acts on particles i.

ξ^n : Coefficients giving the contribution to the heat flux from the generalized electric field $\vec{D}$.

φ^n : Thermal diffusion coefficients.

ψ : Logarithmic cut-off factor.

θ^n : Thermal conduction coefficients.

σ^n : Coefficients of electric conductivity.

ρ_i : Density of particles i.

ρ : Total density.

τ : A collision time for electrons.

ω: The gyromagnetic frequency for electrons.

Index

Balescu, 73, 75, 76, 105, 116
barycentric velocity, 15
Benedicks effect, 8, 65, 66, 114, 120
Boltzmann, 9, 18, 75, 129, 134
Boltzmann equation, 9, 13, 18, 19, 20, 107, 119, 133
Boltzmann's constant, 6, 163
Braginski, 63, 75, 105, 133

chaotic velocity, 15
Chapman, 25, 78
characteristic hydrodynamical length, 109
charge current, 15, 65, 108
CGS, 76
CIT, 5, 8, 10
Clausius, 6, 7
collision integrals, 145, 159
conservation equations, 18
Coulomb, 14, 145, 151
Cowling, 25, 78
cross section, 14
Curie's principle, 93

Dalton's law, 120
Davison, 51
Davison's function, 100
Debye distance, 151
Debye-Hückel, 14
Debye's length, 62, 141
diffusion effects, 41
diffusion flux, 15
diffusive force, 28, 77
distribution function, 13
Donder, T. de, 6
Dorn, 69, 77
Dufour effect, 8, 29, 47, 63, 69, 133

electrical conductivity, 44, 67, 110
entropy flux, 6
Euler, 26, 27, 119, 120

Fickian diffusion, 62, 67, 69
flow of heat, 45
Fourier's equation, 63

Grad, 75
Groot and Mazur, 78, 102
gyromagnetic radius, 151

H Theorem, 18
Hall's effect, 62
heat flow vector, 7
heat flux, 61
Hilbert-Chapman-Enskog, 26, 75
hydrodynamic velocity, 15

internal energy density, 17

Kelvin, 9
Klein's inequality, 19
Knudsen, 75

Landau-Fokker-Planck, 9, 75
Larmor, 18, 159
LEA, 5, 6, 7
Levi-Civita tensor, 85
LIT, 104, 112
local energy density, 6
local entropy, 6
local particles density, 6, 14
logarithmic function, 62, 105

Marshall, 76, 77, 78
mass conservation, 16
mass conservation equation, 108
mass current, 41, 108
mass flux, 8
Maxwell-Enskog, 16
Maxwell distribution, 19, 21
Maxwell's equations, 14, 74, 107, 108, 109, 118, 120, 134
Meixner, 6
momentum, 6
momentum current, 108
momentum flow, 7

Navier-Newton, 112
Navier-Stokes-Fourier, 7, 107, 109
numerical charge density, 15

Ohm's coefficient, 111, 115
Onsager, 9, 10, 29
Onsager matrix, 136
Onsagers' reciprocity theorem, 9, 104, 133, 135
ORT, 9

particle density, 6

Righi-Leduc effect, 47, 62, 67, 68, 76
Rutherford, 146

single particle distribution functions, 13
Sonine (Laguerre) polynomials, 35, 94, 137
Soret effect, 29, 43, 133
Spitzer-Braginski calculations, 63, 75, 133
stress tensor, 93
subsidiary conditions, 28

thermal conductivity, 46, 64
thermoelectric coefficient, 67
Thomson coefficients, 67, 68, 69, 77, 111
Thomsons' thermoelectric effect, 8, 62, 111
total mass density, 13

MIX
Papier aus verantwortungsvollen Quellen
Paper from responsible sources
FSC® C105338

If you have any concerns about our products,
you can contact us on
ProductSafety@springernature.com

In case Publisher is established outside the EU,
the EU authorized representative is:
Springer Nature Customer Service Center GmbH
Europaplatz 3, 69115 Heidelberg, Germany

Printed by Libri Plureos GmbH
in Hamburg, Germany